DEDICATED TO THE

Corona Disease Pandemic Warriors

WHO FOUGHT BRAVELY TOSAVE LIFE OF THE OTHERS DURING THE CRISIS

# GUIDE TO IMPLEMENT INTEGRATED MANAGEMENT SYSTEM
# ( ISO9000+ISO14000+ISO45000)

---

By

**Rajendra S. Patil**

**ACE QUALITY CONSULTANTS**

Fortune World 1, Second Floor, Near ICICI Bank, Jaysingpur

Dist- Kolhapur, Maharashtra, India

acequalityconsultants@gmail.com

**25/4/2021**

## ABOUT THE BOOK

The book is written keeping in view of requirements of implementation of Integrated Management system which is based on ISO 9000, ISO 14000 and ISO 45000 standards,. This book is the outcome of my experience in this field for more than 28 years.

This book covers basic requirements of integrated management system in the form of Simple procedures, which are practically proven and which can be used to implement integrated management system.

The book contains two sections, first section is devoted to the requirements of the integrated management system, and all the requirements of the three standards are covered in this section.

Second section consists of the precautions to be taken while performing work in the organisation during the COVID 19 pandemic. This section includes practical recommendations to organizations and workers on how to manage risks arising out of pandemic and is suitable for organizations resuming operations, those that have been operational throughout the pandemic, and those that are starting operations.

Readers are advised not to use the procedure described in this book as it is, but make necessary changes as per your organizations suitability and effectiveness.

For further details readers are advised to refer the relevant ISO standards

Suggestions are welcome, and can be mailed to,

rajendrapatil295@gmail.com.9822119020

# CONTENTS

## Section 1. IMS REQUIREMENTS

# 1. CONTEXT OF THE ORGANISATION

**General**

Context of the organization is the business environment wherein the organization carries out all its activities. The internal and external issues need to be considered while establishing and implementing the OHSAS system.

The internal issues may include one or more of the following,

- Organization structure, authorities and responsibilities of individual
- Organizational policies and objectives, decision making process
- Relations with the contracted organizations'
- Labour relations
- Working conditions and environment
- Communication channels
- Standard adopted by the organization
- Working hours

The external issues may include one or more of the following

- New Competitors, suppliers, contractors, new laws, technology etc.
- Organizations relations with the above mentioned external parties
- Technical know how about new products and their effects on health and safety of employees
- Social, political, technological, financial and natural surroundings, international, national or Regional market competition.

### 1.1 External and Internal Issues

Determine the external and internal issues that are relevant to its purpose and strategic direction and that affect its ability to achieve the intended results of the IMS.

GIVE consideration to the following,

- Positive and negative factors or conditions.
- External context and issues, such as legal, regulatory, technological, competitive, cultural, social, political and economic environments.
- Internal context and issues, such as values, culture, organisation structure, knowledge and performance of the business.
- Determination and requirements of the needs and expectations of interested parties relevant to the IMS.
- Authority and ability to exercise control and influence.
- Activities, products and services relevant to the business.
- Retain documented information as evidence to support that the context of the organisation has been taken into account in the IMS.

### 1.2 Interested Parties

- Understand needs and expectations of the interested parties, monitor and review the information about these interested parties and their relevant requirements.

Example –

| Interested Party | Needs and Expectations |
| --- | --- |
| Owners/Shareholders | • Less cost of poor quality<br>• Efficient and efficient workforce<br>• Have a growing business that provides profit.<br>• Be well governed and well managed.<br>• Want staff to enjoy their work, be challenged, perform their job competently and meet the company and customer requirements. |
| Customers | • 100% on time delivery<br>• Conformance to quality and cost<br>• Value for money.<br>• A simple solution that manages compliance easier.<br>• Implementation of the product in-line with customer expectations.<br>• Receive responsive support.<br>• Delivery of free content to educate around compliance. |
| Suppliers/Contractors | • Long term business<br>• Ongoing and secure work.<br>• To be paid on time.<br>• Clear understanding of requirements.<br>• Constructive feedback.<br>• Want to provide services/products to a reliable, reputable and financially viable business |
| Partners | • Make them more financially secure through additional revenue<br>• Enable them to change their business model from |

| | |
|---|---|
| | hour-based to value based income. |

| | |
|---|---|
| | • Want a solution that they can sell, promote and support that will assist their client's to manage compliance.<br>• Provide great support and knowledge to help them support their customers. |
| Employees | • Recognition / rewards<br>• Empowerment<br>• Job security.<br>• Salary for work performed.<br>• Flexible work hours.<br>• Clear understanding of their role and responsibilities.<br>• Able to raise issues of concern and provide constructive feedback.<br>• Good, friendly work environment.<br>• To feel valued and appreciated.<br>• Opportunities for personal development. |
| Regulators | • To meet the required laws and regulations.<br>• To submit all tax obligations accurately and on time.<br>• To maintain high standards of corporate governance. |
| Community | • Good corporate citizen.<br>• Diversity of employees |

**1.3 SWOT Analysis**

Carry out strength-weakness-opportunity-threat analysis (SWOT) in order to sustain in the competitive environment in order to determine key business strategies.

| Strengths | Weaknesses |
| --- | --- |
| • Provider of a great quality product.<br>• Provider of great support for the product.<br>• Responsive development to market requirements.<br>• Responsive to identified software issues.<br>• Depth of knowledge of buyer's persona.<br>• Regular delivery of free content.<br>• Low client turnover relative to the industry.<br>• Quick deployment of product post sales.<br>• Deep knowledge of customer's pain<br>• Adaptable, responsive and able to make decisions.<br>• Flexible to meet a wide range of customer service issues.<br>• Open to suggestions to improving the product<br>• Owners have recognised the need to have external expertise to grow the business.<br>• Looking at ways of improving the business. | • Identification of good partners to meet our standards/ requirements.<br>• Managing and review partner performance<br>• Too operational and not strategic enough for partners<br>• Reliance on key employees within the business.<br>• Time poor in a few key areas<br>• Don't have strong relationships with industry players<br>• Measurable marketing outcomes based on known starting points |

| Opportunities | Threats |
|---|---|
| • Changes to standards in our core markets: ISO 9001, ISO 14001, ISO 45001, H&S Act, Food Safety.<br>• New technologies<br>• Partnering with other solutions: Software and Hardware<br>• New focused markets.<br>• Certification to ISO 9001, ISO 45000,ISO 14000 will open up other market opportunities through the marketing of the process.<br>• More marketing via additional platforms<br>• To educate industry in compliance. | • Global Competition<br>• Technology<br>• Changes in government policies and statutory requirements |

### 1.4 Key Business Strategies

| Strategy Description | Action plan for achievement |
|---|---|
| Develop business processes to accommodate the expected growth. | Develop and implement business processes that are suitable for the business.<br>Achieve certification to ISO 9001.<br>Transfer of knowledge to partners and employees for all key processes<br>Use technology to manage as many processes as appropriate |
| Improve the efficiency and effectiveness of the core processes | Identify the core processes (i.e. development and release, sales, marketing, implementation, support)<br>Identify new ways (e.g. lean techniques) of |

| | |
|---|---|
| | doing the core processes<br>Update and embed the core processes to ensure knowledge is retained |
| Personnel to be capable of delivering the growth for the business | Key leadership personnel to be capable of leading and managing their staff.<br>Competency gaps to be identified by leadership personnel to be assessed as competent for their role Personnel to receive training for the role<br>Personnel to receive appropriate experience to do the role |
| Grow market share in all markets | Identify and train new partners<br>Continuously review partner performance<br>Identify changes to legislation, standards and regulation Identify key market verticals in each jurisdiction<br>Increase the number of qualified lead by creating more content and deliver across multiple channels<br>Improve the sales conversion rate from qualified leads to sale |

**1.5 Management Representative**

Appoint management representative for co-coordinating all IMS activities

Management representative ensures that,

a. IMS is established, implemented and maintained in accordance with the requirements of ISO 9001:2015, ISO 14001:2015 and ISO45001:2018.
b. IMS processes are delivering their intended outputs.

c. Promotion of customer focus throughout the company.

d. Integrity of the IMS is maintained when changes to the IMS are planned and implemented.

Management representative reports on the performance of the IMS to top management for review and as a basis for improvement.

## 1.6 IMS Structure

**Interaction of Processes in the IMS**

The company's IMS should comply with:

- ISO 9001:2015,
- ISO 14001:2015
- ISO 45001:2018

While establishing the IMS system, prepare following levels of documented information

Policies: Policies are documents that demonstrate the overall commitment to improving quality performance and are authorised by the Management Team.

- System procedures: high-level procedures that define the activities that are to be fulfilled to ensure that the IMS that complies with standards.
- Data base workflows, operational procedures and work instructions. Control and operational procedures:
  - Meet customers' requirements.
  - Provide supplementary guidance and instructions to support the intent of the IMS.
  - Ensure that the requirements of the IMS will be adequately addressed within the organisation.
- Forms, registers and records are evidence to prove the IMS is operational.

- Written procedures and forms associated with the process are prepared and utillised. They include the following:

| Activity/processes | Actions |
|---|---|
| Accident/Incident | Controls all near miss, accident, incident or injury reporting, investigation, corrective action and injury management. The workflow manages the process. |
| Audit/Inspection | Controls the audit process, from scheduling through to audit reports. Its captures all the corrective actions in the Improvement module. |
| Compliance | Controls all the reporting and investigation of legal, regulatory and standard registers in a seamless workflow. |
| Documents | Controls and maintains the approval, publishing and authorisation of all documentation within the data base with electronic signatures. |
| Event Management | Controls scheduled tasks to ensure critical activities can be managed and monitored. It includes email reminders and alerts when events are due or overdue. |
| Human Resources | Controls all a comprehensive central database of employee details that links people with their skills and positions. |

| Improvement | Controls all reporting, investigation and corrective actions for customer complaints, audit findings, internal issues or non-<br>Conformances. The workflow manages the process. |
|---|---|
| Plant/Equipment | Controls the maintenance of plant and equipment with email reminders and alerts when maintenance is due or overdue. |

| Risk Management | Controls identified and managed risks. The workflow manages the<br>process |
|---|---|
| Supplier and Contractor Management | Controls all external suppliers and contractor's details and performance is captured and reviewed. |

# 2. Leadership and Commitment

Purpose -- Document and establish the procedure which shows how company demonstrates leadership and commitment to its IMS.

## 2.1 Procedure

1. Top management will take responsibility for the effectiveness of the IMS and will demonstrate their commitment to the IMS by:
    a. Defining roles, allocating responsibilities and accountabilities, and delegating authorities, to facilitate effective IMS management.
    b. Roles and Responsibilities are documented in Leadership - Organisation Roles, Responsibilities and Authorities and through position descriptions, and IMS procedures where applicable. Ensuring:
        i. That relevant policies and objectives are established for the IMS and that these are aligned with the context and strategic direction.
        ii. The integration of the IMS requirements into the organisation's business processes.
        iii. Those resources needed for the IMS are available.
        iv. The IMS achieves its intended results.
        v. The process approach and risk based thinking is promoted. Communicating the importance of effective IMS management and of conforming to the IMS requirements.
        vi. Engaging, directing and supporting personnel to contribute to the effectiveness of the IMS.
        vii. Improvement is promoted.
        viii. Other relevant management roles are supported to demonstrate their leadership as it applies to their areas of responsibility.

2. Top management is committed to our customers and enhancing customer satisfaction. This commitment is demonstrated by:
    a. Ensuring that applicable customer and statutory requirements are determined, understood and met throughout the business.
    b. Ensuring the risks and opportunities that can affect conformity of products and services and the ability to enhance customer satisfaction are determined and addressed.
    c. Exercising due care with our customer's property (data) whilst it is under the control of the company.
    d. Monitoring customer's perceptions of the degree to which their needs and expectations have been fulfilled.
3. The key aspects of the customer information and data generated through the effective implementation of the IMS processes are collected and collated by the Management Representative and presented at each Management meeting.

The organisation should demonstrate its commitment to providing and delivering the customer great product, great support and great marketing to exceed customers satisfaction and make customer happy.

In order to achieve quality policy, objectives are also need to be established by means of which quality policy will be achieved

## 2.2 Quality, Environmental, Health and Safety (QEHS) Policy.

QEHS policy is a top level document that reflects the commitment of top management of a company to its intention of implementing the measures for achieving the quality and HSE targets and objectives. Above the policy there are two more documents which most organizations keep i.e. Mission and vision of the company. In additions a company may define its business ethics. Below is a sample policy which can be useful for you to develop the QEHS policy for your company

**Quality, Environmental, Health and Safety Policy**

It is the Policy of XYZ organisation that the highest priority is given to all aspects of quality, health, safety and the environment in all its business activities and that the highest standards are implemented in all Operating and service Departments. The company's quality, health, safety and environmental procedures are linked within an Integrated Management System. The System encompasses compliance with client requirements, delivery of exceptional service, elimination of costs with no added value, proactive waste management, and prevention of pollution and minimization of any adverse impact on environment. It also embraces a safe place of work, safe systems of work and a healthy working environment to prevent injury, ill health or damage. [COMPANY NAME] considers that duties to comply with ordinances, regulations and other stated requirements should be regarded as a minimum requirement only.

**2.3 Objectives --**

1. Focus on customer satisfaction. Ensure the highest standards of quality, health, safety and the lowest environmental Elimination of Major accidents and incidents. Systematic reduction of minor accidents and incidents.
2. Providing and maintaining a culture of free and honest incident and accident reporting.
3. Prevent pollution and implement resource conservation as critical management considerations.
4. Establish clear quality, health, safety and environmental objectives and targets.

Provide all necessary resources to ensure a safe place of work and safe systems of work.

5. Maintaining a focus on occupational health and health surveillance, where required. 6. Provide an effective system of work related injury and illness rehabilitation.

Provide all necessary information, training and supervision to all its personnel and subcontractors.

7. Continually improve through regular performance reviews.
8. Provide customers with a quality product for the management of their compliance needs.

9. Provide customers with free content, information and industry insight to improve their compliance knowledge.
10. Provide timely and accurate support to customers
11. Listen to customers when developing and enhancing the product.
12. Provide an environment where staff can grow and learn new skills.
13. Setting objectives
14. Documenting plans
15. Reviewing

THIS can be achieved by,

- Training employees
- Training Partners
- Investing in resources
- Investigating in new technologies

The organisation must identify environmental management as one of its highest corporate priorities. The organisation should develop policies, programmes and practices to reduce risk to the environment and the organisation and conduct business activities in an environmentally sound manner.

The organisation should commit itself to environmental management system and exercise following,

- Integrate its environmental policies and procedures fully into all business activities as a critical element,
- Comply with all environmental legislation, standards and contract requirements that are applicable to the company's operation,
- Continually improve its environmental performance and prevention of environment impact and taking into account current best practice, technological advances, customer and community needs, educate, train and promote employees to work in an environmentally responsible manner,
- Complete environmental assessments for aspects and impacts of all new

activities that the company may undertake, promote, develop and design services, facilities, equipment and work practices that have the least environmental impact, taking into account the efficient use of energy and materials, the sustainable use of renewable resources and the responsible disposal of waste, thereby minimising any serious or irreversible environmental degradation,

- Promote and encourage the adoption of these principals by suppliers and contractors acting on behalf of the organisation,
- Develop, implement and maintain emergency preparedness plans,
- Foster openness and dialogue with both employees and the public, encouraging them to respond with their concerns or improvement ideas within the scope of the organisation's operations and maintain a set of environmental objectives and targets that are monitored through the management review process to ensure effectiveness.

Management will:

- Set health and safety objectives and performance criteria for all managers and work areas Annually review health and safety objectives and managers' performance
- Encourage accurate and timely reporting and recording of all incidents and injuries
- Investigate all reported incidents and injuries to identify all contributing factors and, where appropriate, formulate plans for corrective action
- Actively encourage the early reporting of any pain or discomfort
- Provide treatment and rehabilitation plans that ensure a safe, early and durable return to work
- Identify all existing and new hazards and take all practicable steps to eliminate or minimise the exposure to any hazards
- Ensure that all employees are made aware of the hazards in their work areas and are adequately trained so they can carry out their duties in a safe manner

- Encourage employee consultation and participation in all health and safety matters Enable employees to elect health and safety representatives
- Ensure that all contractors and subcontractors are actively managing health and safety for themselves and their employees
- Promote a system of continuous improvement, including annual reviews of policies and procedures
- Meet legal obligations as specified in the legislation, codes of practice and any relevant standards or guidelines.

Every manager, supervisor or foreman is accountable to the employer for the health and safety of employees working under their direction.

Each employee is expected to help maintain a safe and healthy workplace through:

- Share in the commitment to health and safety.
- Following all safe work procedures, rules and instructions
- Properly using all safety equipment and clothing provided
- Reporting early any pain or discomfort
- Taking an active role in the company's treatment and rehabilitation plan, for their 'early and durable return to work'
- Reporting all incidents, injuries and hazards to the appropriate person.

The Health and Safety Committee includes representatives from senior management and union and elected health and safety representatives. The Committee is responsible for implementing, monitoring, reviewing and planning health and safety policies, systems and practices.

It is the duty of every employee of the organisation to implement the Management System in the performance of their duties and to ensure that this policy is supported and maintained. This policy will be communicated openly to all employees, customers, suppliers, and subcontractors. It will also be made available to other interested parties whenever requested. The Managing Director / General Manager of the organisation shall be responsible for the overall implementation of the Quality, Health, Safety and Environmental Management System while the Management System Leader shall be responsible for the overall management of the Integrated Management System.

## 2.4 Organisation Roles, Responsibilities and Authorities

**Purpose and Scope**

To describe the responsibilities , authorities for the IMS and to define the organisation structure for the effective operation of the IMS.

Requirements

1. The responsibility, accountability and authority of all personnel involved in the IMS need to be defined, documented and communicated in order to facilitate effective IMS. This is to include any responsibilities and accountability that is imposed by legislation.
2. Responsibilities, accountabilities and authorities are documented in position descriptions and throughout the IMS.
3. Where suppliers are involved, their responsibilities and accountabilities need to be clarified and documented by the responsible employee with authority.
4. All employees and Suppliers will comply with their

responsibilities. The top Management need to:

1. Ensure organisation-wide compliance to the IMS.
2. Appoint the IMS Management Representative.
3. Ensure that the assigned roles, responsibilities and authorities are communicated and understood.
4. Communicate the importance of meeting customer, statutory and regulatory requirements.
5. Establish appropriate policies that include a commitment to continual improvement of the IMS.
6. Establish IMS objectives.
7. Ensure that all employees are aware of:

a. Policies.
b. Current IMS objectives, targets and plans.
c. The importance of compliance with the IMS.
d. Their contribution to the effectiveness of the IMS, including the benefits of improved performance.
e. Potential consequences of non-compliance with the IMS requirements.

8. Hold people accountable for carrying out assigned responsibilities and the results delivered.
9. Make resources available.
10. Participate in IMS meetings including the Management Review.
11. Utilise data base for the effective control of the IMS.
12. Actively promote and participate in IMS

initiatives. The Management Representative is to:

1. Ensure that the:
   a. IMS is established implemented and maintained in accordance with the requirements of the standards.
   b. IMS processes are delivering their intended outputs.
   c. Promotion of customer focus throughout the company.
   d. Integrity of the IMS is maintained when changes to the IMS are planned and implemented.
2. Report on the performance of the IMS for review and as a basis for continual improvement.
3. Perform the role of Administrator which has the authority to ensure access rights in the IMS, for individuals, are in-line with their levels of authorities and responsibility in the organisation.
4. Monitor, communicate and incorporate changes in the legal and other requirements in the IMS.
5. Communicate amendments to the IMS.
6. Advise and provide guidance to ensure compliance to the IMS is maintained.
7. Provide guidance in developing action plans and conducting management

system reviews.

8. Ensure that audits and inspections are conducted in accordance with the schedule.
9. Ensure that Mango is effectively utilised to administer and control the IMS.
10. Provide and or arrange for ongoing training and coaching to personnel with respect to IMS matters.
11. Coordinate and participate in IMS meetings including the Management Review.
12. Publish and control all IMS documents.
13. Actively promote and participate in IMS initiatives.
14. Coordinate and administer arrangements with the

certification agency. Employees are to:

1. Ensure that the IMS is effectively implemented and maintained within their area of responsibility.
2. Actively encourage all personnel to contribute towards the continual improvement of the IMS.
3. Incorporate the IMS as part of site and departmental inspections and reviews.
4. Determine and escalate the need for resource requirements for the effective operation of the IMS.
5. Participate in IMS meetings including the Management Review.
6. Utilise company data base for the effective control of the IMS.
7. Actively promote and participate in IMS initiatives.
8. Promptly report any unsafe working conditions, faulty equipment, hazards/risks, injuries or incidents

Suppliers and Contractors are to:

1. Comply with the requirements of the IMS and participate in IMS promotions.
2. Promptly report any unsafe working conditions, faulty equipment, hazards/risks, injuries or incidents

**2.5 Organisation Structure**

1. The Company recognises that the structure of the organisation needs to constantly evolve in order to meet the changing needs of clients, the market and compliance obligations.
2. The Management Team are responsible for ensuring the structure of the organisation is appropriate to the current business needs and will ensure that the organisation chart is regularly reviewed and maintained.

# 3. Planning

## 3.1 Actions to Address Risks and Opportunities

**Purpose and Scope**

To describe the manner in which the company identifies and manages the risks and opportunities within the business.

**Procedure**

1. The company is committed to identifying and addressing relevant risks and opportunities as a means for:
   a. Increasing the effectiveness of the IMS.
   b. Improving performance.
   c. Preventing or mitigating negative effects.
2. When undertaking risk management activities the consideration should be given to the:
   a. Positive and negative factors or conditions.
   b. External context and issues, such as legal, regulatory, technological, competitive, cultural, social, political and economic environments.
   c. Internal context and issues, such as values, culture, organisation structure, knowledge and performance of the business.
   d. Determination of the requirements and needs and expectations of interested parties relevant to the IMS.
   e. Authority and ability to exercise control and influence.
   f. Activities, products and services relevant to the business.
3. Organisation may adopt any or a combination of the following risk options:
   a. Avoid the risk.
   b. Eliminate the risk source.
   c. Take the risk to pursue an opportunity.
   d. Change the likelihood or consequences of the risk.
   e. Share the risk.

f. Retain the risk by informed decision.

4. Opportunities identified by the company may lead to:
    a. Adoption of new and improved processes.
    b. Launching new products or services.
    c. Pursuing new markets.
    d. Utilising new technology.
    e. Improved ways of addressing customer needs.
5. Organisation can manage risk and opportunities as follows:
    a. Through ongoing effective leadership and commitment to the IMS.
    b. Manage business and quality risks and opportunities in the Board meeting and the Management meetings.
    c. Through the effective management and control of suppliers and contractors.
    d. Through the effective training of personnel to ensure they are competent to perform relevant tasks safely.
    e. By monitoring, measurement and review of relevant processes and outputs.

### 3.2 Legal and OtherRequirements

**Purpose and Scope**

To describe how the organisation ensures that it has identified, complies with and verifies compliance with all of the relevant legislative, regulatory and other requirements that apply to the activities conducted by the its employees and contractors within its operations.

**Procedure**

1. The company is to ensure that all relevant legislative and other requirements are identified.
2. Legislative and other requirements may include, but are not limited to:
    a. Acts and Regulations.

b. Codes of Practice.
c. Guidelines.
d. Standards.
e. Agreements with clients, communities or public authorities.
f. Corporate requirements.
g. Industry standards or codes.
h. Voluntary commitments.

3. Details of all relevant legislative and other requirements are to be contained within the Compliance Module. These will include mitigations and control methods. The verification of compliance will be reviewed by the Board.
4. The Management Team are to ensure that where possible, they are notified of changes and/or additions to legal and other requirements as those changes occur.
5. The means of ensuring notification of changes and/or additions may include:
   a. Agreements with external legal or consulting organisations to monitor and advice of any changes.
   b. Registering with Standards New Zealand.
   c. Advice from employer or industry associations.
6. When changes and/or additions occur they are to be included in the Compliance data base and the means of verifying compliance is to be defined as previously described.
7. A review of the Compliance data base will be conducted as per the annual work plan in the Board meeting. These will include:
   a. Confirm that all updates to applicable legal and other requirements have been captured and included.
   b. Confirm that the means of ensuring and verifying compliance are appropriate.
8. The company is to ensure that all changes, additions and updates to the Compliance data base are communicated to relevant employees, contractors and other stakeholders.

## Means of Ensuring Compliance

1. Once the application of a particular requirement has been defined, the means of how compliance to the requirement is going to be ensured is to be established by the company, in consultation with appropriate personnel.
2. Various means of ensuring compliance are available and include, but are not limited to the following:
    a. Policies and/or procedures being established documented and implemented.
    b. Training being provided.
    c. Engineered solutions being implemented.
    d. Instructional signs being displayed.
3. Details of the means of ensuring compliance are to be entered into the Legal and Other Requirements Register in the "Means of Ensuring Compliance" column alongside the corresponding requirement in the Compliance Module.

## Means of Verifying Compliance

1. Once the means of ensuring compliance has been determined, the means of how compliance to each requirement is to be verified on a continuous basis is to be established by the company, in consultation with appropriate personnel.
2. Various means of verifying compliance are available and include, but are not limited to the following:
    a. Internal auditing. (To verify compliance to the corresponding policies and/or procedures).
    b. Periodic workplace inspections.
    c. Periodic review of records.

3. Details of the means of ensuring compliance are to be entered into the Legal and Other Requirements Register in the "Means of Ensuring Compliance" column .The Management Team are to ensure that where possible, they are notified of changes and/or additions to legal and other requirements as those changes occur.

4The means of ensuring notification of changes and/or additions may include:

a. Agreements with external legal or consulting organisations to monitor any advice of changes.
b. Registering with Standards or Government organisations.
c. Advice from employer or industry associations.

5 When changes and/or additions occur they are to be included in the Legal and Other Requirements Register and the means of ensuring and verifying compliance is to be defined as previously described.

6.On an annual basis, usually in Q4 each year, the Management Team, in consultation with appropriate personnel, is to coordinate a full review and update of the Legal and Other Requirements Register in order to:

- Confirm that all updates to applicable legal and other requirements have been captured and included.
- Confirm that the means of ensuring and verifying compliance are appropriate.

7 .The Management Team is to ensure that all changes, additions and updates to the Legal and Other Requirements Register are:

f. Tabled at management review and other relevant meetings.
g. Communicated to relevant employees, contractors and other stakeholders.

**3.3 Objectives, Targets andPlans**

**Purpose and Scope**

To define the processes for establishing measurable IMS objectives and targets, for establishing plans to achieve those objectives and targets and for periodically monitoring performance in achieving each objective and target.

**Procedure**

1. The company will establish measurable objectives and targets in relation to its IMS performance.
2. The established objectives and targets must be:
    a. Consistent with the applicable policies.
    b. Measurable.
    c. Monitored and updated.
    d. Effectively communicated to relevant parties.
3. When establishing, reviewing and updating measurable objectives and targets, consideration is to be given to:
    a. Health and safety hazards/risks.
    b. Significant environmental aspects and risks/opportunities.
    c. Significant business or quality risks/opportunities.
    d. Technological, financial and Operational and business requirements.
    e. Products and services provided to customers.
    f. The enhancement of customer satisfaction.
    g. Views of stakeholders.
    h. Legal and other requirements.
4. Once measurable objectives and targets have been established, plans for achieving those measurable objectives and targets are to be established.
5. Performance in achieving each measurable objective and target is to be periodically monitored during Management Review meetings.

# 4. SUPPORT

## 4.1 Resources and Infrastructure

### Purpose and Scope

To describe how the resources and infrastructure required to establish, implement, maintain and continually improve the effectiveness of the IMS and business operations are identified, provided and maintained.

### Procedure

1. Resources include human resources and infrastructure, technology and financial resources.
2. The infrastructure and work environment needed to achieve conformity to product requirements is to be determined, provided, managed and maintained. This can include, as applicable:
    a. Buildings and associated utilities.
    b. Equipment including hardware and software.
    c. Information and communication technology.
3. The Management Team will provide the organisational infrastructure, technology and financial resources. They are to review the adequacy of the resources.

    As new technology becomes available, the possibility of introducing it to improve the IMS is to be considered.
4. The Management Representative is to identify the resources required to establish and maintain the IMS.
5. The Management Team is to priorities the financial resources available and allocates them to the various departments to provide the resources needed.
6. Each Department is to identify the resources required and to provide adequate support when planning work. They are to identify the infrastructure needed to implement and continually improve the IMS and meet requirements. The infrastructure to be considered could include, but

is not limited to:

  a. Buildings and workspace.
  b. Hardware and Software.
  c. IT requirements.
  d. Communications.

7. The Management Team will determine and maintain an appropriate work environment needed to achieve conformity to the product or service requirements.

## 4.2 Plant and Equipment

1. Details of equipment used by employees are recorded in the PPE/Item data register and on the asset register.
2. All repairs, must be carried out:
   a. In accordance with any regulatory and the original manufacturer's requirements.
   b. By appropriately trained, qualified, competent and experienced personnel.
   c. All records of maintenance are recorded on the suppliers invoice

## 4.3 Training, Competency and Knowledge Management

**Purpose and Scope**

To ensure all relevant personnel are adequately trained, competent and informed in accordance with their position and IMS requirements.

**Procedure**

Commencement and Induction of New Employees:

1. Employee setup in Employee history sheet

1. Before a new employee commences work, the employee's manager is to arrange for induction training in accordance with the induction checklist.
2. During the induction any training needs will be identified and logged in

Employee history sheet

3. Once completed the induction checklist must be signed and dated by both the new employee and the employees' manager.
4. A record of the induction is to be maintained in employee history card. .

**Initial Employee Assessment:**

1. The employee's manager assesses the employees' competency against the skill set that has been established within the Skills/Qualifications matrix
2. The employee and the manager agree current competency and future training needs.
3. The Skills/Qualifications matrix for that employee is updated by the employees' manager or delegate. Any supporting records are also maintained in training history sheet.
4. The next review date for any further assessment of the employee's competency and training needs is to be scheduled in the training calendar and same shall be used as a reference for the next review.
5. Scheduled training is also able to be captured within the Events Management Module, if necessary.
6. The employee's manager is to ensure that training identified is undertaken, and whilst under training the employee is appropriately supervised, as may be required.

Further Assessment of Employee Competency and Training Needs

1. The employees manager is responsible for conducting:
   a. 90 day performance reviews.
   b. Ongoing performance reviews.
   c. Further assessment of employee competency and training needs.
2. The further assessments of employee competency and training needs are conducted and involves the following steps:
   a. Upon notification from, the manager will conduct an assessment of the employee.
   b. The employee and manager agree current competency, review

training undertaken during the previous year and evaluate the effectiveness of it and decide on future training needs.

c. The Skills/Qualifications record for that employee is updated by the manager or delegate. Any supporting records are also loaded into Employee history card at this time.
d. The next review date for assessment of the employee's competency and training needs is to be scheduled in the history card
e. Scheduled training is also able to be captured within the training calendar
f. The manager is to ensure that training identified is undertaken, and whilst under training the employee is appropriately supervised.

**4.4 Induction of Suppliers**

1. Relevant suppliers must be inducted prior to commencement of work in accordance with the applicable induction checklist. Records of the induction are to be retained.
2. During the induction they will be advised of any potential hazards/risks together with information about required control measures and emergency procedures.
3. The induction is to also cover (as applicable):
    a. Quality Policies.
    b. Current IMS objectives, targets and plans.
    c. The importance of compliance with the IMS.
    d. Their contribution to the effectiveness of the IMS, including the benefits of improved performance.
    e. Potential consequences of non-compliance with the IMS

requirements. Training Providers

1. In-house training is to be conducted by appropriately skilled and competent trainers with relevant experience, depending upon the subject matter.
2. Training may be performed by suitably trained, qualified and experienced

external service providers.

### 4.5 Knowledge Management

Knowledge can be captured by one or more of the following tools,

- Monthly Company meetings
- Weekly production meetings
- Weekly quality and marketing meetings
- Use of accounting system

Capture this knowledge in each of these tools and share it amongst the company to ensure the knowledge is used in giving the customer value. Review the effectiveness and efficiency of these sources at defined intervals in the company's Management Review.

### 4.6 Communication, Consultation and Awareness

**Purpose and Scope**

This outlines the framework for communication and consultation with employees, contractors and external parties in relation to IMS issues and initiatives.

The main objectives are to ensure personnel at all levels and functions;

Are aware of IMS requirements and are effectively involved in the development, implementation and review of policies and procedures.

- Consulted when there are any changes that affect the workplace and or IMS systems.

**Procedure**

Communication of IMS Information with the Board:

The IMS and legal requirements are communicated and discussed at the board level. The BOD minutes record what items have been discussed and actions to be done.

Where required, actions will be assigned in the monthly meetings.

Communication and Awareness of IMS Information with Internal Parties:

1. The IMS communication and consultation processes will occur at the monthly company meeting run by the Director/s and attended by all employees.
2. The meeting will have an agenda that includes, but not limited to:
    a. Quality Policies, Objectives, targets and plans.
    b. Customer complaints and actions taken
    c. Risks and Issues
    d. Marketing, Sales, Support and Development improvement and performance.
    e. Audits/Process Improvement
3. An email will notify the owner when meetings are due and will be signed off the event including relevant evidence attached.

The company has ad-hoc meetings support the consultation processes:

| **Forum** | **Attended by** |
| --- | --- |
| Quality meeting | Q A staff+ Management |
| Production Meeting | Production Team + Management |
| Marketing Meeting | Marketing Department |
| Support | Support Team + Management |
| Development Meeting | Development Team |
| | |

Communication and Awareness of IMS Information to External Parties:

The company will communicate information externally about its IMS performance based on their enquiry.

## 4.7 Documented Information and Control of Documents

### Purpose and Scope

To describe the methods to control and manage documented information critical to the IMS.

### Procedure

1. Documented information includes manuals, policies, procedures, work instructions, forms, registers, flow charts, records and other IMS document requirements.
2. The Management Representative is responsible for ensuring that all IMS documented information is effectively controlled.
3. All employees are responsible for ensuring they are always up to date with all IMS documented information available in the Documents module.
4. Copies of procedures, policies and other documented information may be printed from Mango, but these printouts will be deemed "uncontrolled".
5. To prevent the unintended use of obsolete documented information, superseded documents are automatically identified and removed from general view through the workflow. Obsolete documents are only able to be accessed by personnel, with the required access levels, through the "History" button in the Documents Module.

Editing, Approval, Publishing and acknowledge of Documents:

The Documents data base manages the following document control activities:

1. Creation and editing

- Approval
- distribution
- Acknowledgment

- Retention of previous version
- Revision numbering
- Notifications

Requests for changes:

1. All requests must be raised through document change note.

Advice of Changes:

1. When a change is made or new document added, personnel are able to be notified by email at the time of publication.
2. Changes can also be communicated via monthly meetings as deemed appropriate.
3. Changes to all IMS documents can be tracked through the Document Change History card

Maintenance of IMS Documents

1. All IMS documents are to be reviewed at least once every three years, revised as necessary and approved for adequacy.
2. This review is to be coordinated by the Management Representative in conjunction with the relevant competent and responsible personnel as determined by the Management Representative at the time of review.

External Documents

1. It is the responsibility of the Management Representative to review, implement and maintain external documents and verify that they remain current.
2. All external documents are verified as current and when necessary have their distribution controlled as described in this procedure. Updates to external documents shall be placed in the appropriate file in the Documents and approved and published in accordance with this procedure.
3. The Management Representative subscribes to relevant external regulators, agencies and bodies who may provide periodic advice of changes to their

specific documents. Upon receiving advice of changes to an external document the Management Representative will incorporate this change in the document and ensure the change is communicated to relevant parties.

Computer Back-Up

1. The Management Team are responsible for ensuring that appropriate arrangements are in place to ensure that a back-up of data stored on the server is carried out on a daily basis.
2. The IMS as documented is backed up automatically by the application. Back-ups are captured each hour within the primary data centre with additional back-ups being captured every eight hours at a secondary data centre.

**4.8 Records Management**

1. All IMS records are retained in document section for as long as the company requirements.
2. All IMS Procedures and Forms are maintained within the Documents section
3. The Management Representative is responsible for the management of records with respect to the IMS.

# 5 OPERATIONS

## 5. 1 design and development planning

Plan and control design and development of product, determine

- Design and development stages
- Review, verification and validation stages in design
- Responsibility and authority for design and development
- Manage interface between different groups involved in design and development
- Update output of planning

Plan, design and development activities, decide design and development stages, review, verification and validation, authorities and responsibilities of personnel with target date.

Record of design and development planning shall be maintained

## DESIGN AND DEVELOPMENT INPUTS

Determine inputs related to product requirements including,

Functional and performance requirements

- Applicable statutory and regulatory requirements
- Information of previous reviews
- Other essential information for design and development
- Review input for adequacy

Organization shall determine design and development inputs. Design and development inputs may be determined from,

- Customer drawing and specifications
- Enquires
- Similar products design

- Customer feedback, customer complaints
- Benchmarking with competitor design
- Review of design and development input shall be carried for following,
- Adequacy of the input
- Conflicting specifications
- Irrelevant specifications

Records of design and development input review must be maintained. format for the same is given

## PRODUCT DESIGN INPUT

Identify, document and review product design input including,

- Customer requirements such as contract review, identification and traceability
- Use of information such – establish processes to deploy information gained from previous design projects, concept analysis, and internal audit etc.
- Targets for product quality, life, reliability

Obtain input from contract review, customer feedback, customer complaints, market survey,
Inputs can also be obtained from competitor's products, benchmarking etc. all documents which are reviewed against which parameter the record of such input and review must be maintained.
For standard products design Input review can be carried out against standard checklist, catalogues etc.

## MANUFACTURING PROCESS DESIGN INPUT

Identify, document and review manufacturing process design input requirements including,

- Product design output data
- Productivity targets, process capability and cost
- Customer requirements if any
- Past development experience

Manufacturing process design includes process finalization, production planning, measurement of processes, process parameter selection; tooling selection etc. input to manufacturing process design must be documented and reviewed. The input to manufacturing process design may be,

- Customer schedule
- Machine, process capability data.
- Product drawing, design
- Past experience of similar types of products

Control special characteristics. Reaction plan must be in place for special characteristics

## DESIGN AND DEVELOPMENT OUTPUTS

Provide design and development output in such a form so as to enable verification against design and development input requirements. Output shall,

- Meet input requirement of d & d
- Provide information for purchase, production and service provision
- Contain or refer product acceptance criteria
- Specify characteristic of product essential for safe and proper use.

Design and development output must be,

- In co relation with design and development input
- In a suitable form
- Provide information for production, purchasing activities

- Include product acceptance criteria
- Specify special characteristics of the product.

Design and development outputs must be approved by designated authority before release.

## PRODUCT DESIGN OUTPUTS

Express output in a form enables to verify against input requirements and shall include,

- Design FMEA, reliability results
- Special characteristics and specifications
- Product error proofing
- Product drawing
- Design review results

Product design output may be,

- Drawing, specifications
- Bill of material
- Purchase orders

Product design output must be expressed in terms that can be verified and validated against product design input requirements.

## MANUFACTURING PROCESS DESIGN OUTPUT

Express manufacturing process design output that can be verified against manufacturing process design input requirements and validated, manufacturing process design shall include,

- Specifications and drawings
- Manufacturing process flow chart
- Manufacturing process FMEA
- Control plan
- Work instruction
- Process approval acceptance criteria
- Data for quality, reliability and maintainability
- Results of error proofing activities as appropriate
- Method of rapid detection and feedback of product/manufacturing process non-conformities

Manufacturing process design output may be,

- Production plans
- Flow charts
- Quality plans, process plans
- Operator instructions
- Failure mode and effects analysis
- Specifications and drawings

There must be correlation between manufacturing process design output and manufacturing process design inputs.

## 5.2 DESIGN AND DEVELOPMENT REVIEW

- Perform d & d review at suitable stages as per planning
- Evaluate ability of the results of d & d to meet requirements
- Identify problems and propose necessary actions
- Involve all functions in d & d review
- Maintain records of d & d review

Design and development review must be carried out at defined interval. D & D review should be carried out by cross-functional team. D& D review shall be carried out to ensure the following,

- Suitability of design
- Compliance with statutory and regulatory requirements
- Compliance with customer requirements
- Consideration of unintended usages
- Problems that could be encountered during further processing
- Interchangeability of pats and use of substitute material.

Records of design and development review must be maintained.

Present the results of d &d reviews, analysis and keep it as input for management review.

**5.3 Design and development verification**

- Verify design as per planned arrangements
- Ensure d&d output meets d&d input requirements
- Maintain records of verification

Design and development verification performed to ensure design output meets design input requirements. Verification may be performed by one or more of the following means,

- Performing alternative calculations
- Comparing with similar past design
- Comparing with competitors design of the same product.

The records of design and development verification must be maintained

## 5.4 design and development validation

- Perform d&d validation in accordance with planned arrangements
- Ensure product meets requirements of specified applications
- Complete validation prior to delivery (where practicable)
- Maintain records of validation

D & D validation shall be performed as per validation plan. D & D validation shall be carried out to ensure that resulting product meets specified requirements of applications. Validation may be performed in-house or at site. Validation may be performed by one or more of the following means,

- Performing trials/ testing as per field applications on actual product in-house
- Performing trials/tests at customers end
- Performing tests on models
- Analysis of field reports

Records of design validation and its results must be maintained.

## CONTROL OF DESIGN AND DEVELOPMENT CHANGES

- Identify d &d changes and maintain records
- Review, verify and validate changes before implementation
- Evaluate effect of changes on constituent parts
- Maintain records of reviews and necessary actions

D &D changes must be suitably identified. Record of D &D changes must be maintained including nature of change, basis of change, revision details etc. D & D changes must be reviewed for following,

- Effects on end product
- Effects on manufacturing processes
- Effects on tooling
- Effects on constituent parts

D & D changes must be performed by same authority who has initially performed D& D reviews. Record of design change review must be maintained.

# 6. PRODUCTION AND SERVICE PROVISION

## 6.1 Control of production and service provision

Plan and carry out production under controlled conditions. Controlled conditions may be,

- Availability of product related information
- Availability of documents like work instruction, process plans
- Availability and use of measuring and monitoring devices
- Implementation of inspection and testing activities
- Implementation of delivery activities.

Production planning must be carried out based on customer schedule/ requirement. Schedule received from customer must be reviewed for quantity and delivery period.

Production plan must be communicated to all concern. Consideration must be given to following during production,

- Availability of all documents that describes the requirements of the product. The documents may be quality plans, process plans, work instructions etc.
- Availability of skilled manpower
- Availability and adequacy of tooling, gauges.
- Provision of safety equipments and working conditions

## 6.2 Work instructions

Prepare and make available work instructions at all locations and for all personnel carrying work affecting quality. Take input to work instruction from quality plan, control plan and product realisation process.

Work instructions must be prepared for all operations/areas where absence of such could affect quality of product. Work instruction must be displayed at workstations. Main purpose of work instruction is to guide workers to perform a particular task. So it is advisable to prepare work instructions in the language which can be easily understood by the workmen i.e. in the mothertounge. Work instructions should cover following,

Use of control plans, FMEA s can be made while preparing work instructions.

- Activities to be performed in sequence.
- Precautions to be taken while performing activities
- Tools other resources to be used for performing activities
- DO s and DON'Ts while performing activities
- Stages of inspection.
- Reaction plan

**6.3 Validation of processes for production and service provision**

Validate production and service provision processes where the resulting output cannot be verified by subsequent monitoring or measurement and deficiencies may become apparent only the product is taken for use. Validation shall demonstrate the ability of these processes to achieve planned results. Establish arrangements for following

- Defined criteria for review and approval of these processes
- Approval of equipment and qualification of personnel
- Use of specific methods and procedures
- Requirements of records and revalidation

The processes, where the output cannot be assured as ok or reject after immediate inspection and testing and the deficiencies in the products come to surface after the product has been taken for use. Such processes are known as special processes. Examples of these process are, welding, heat treatment, powder coating etc.

Ensure the ability of these processes to achieve planned results in order to achieve this validation of such processes must be carried out. Validation should be carried out for,

- Process validation
- Equipment validation
- Operator validation

Process validation shall be carried out by conducting range qualification testing. In this experiment the product is subject to different range of specifications e.g. in welding process the product is welded for different range of currents And the results on product is tested i.e. quality of welding is tested for different current.

The current value at which welding quality is found satisfactory is taken for regular practice. Similar type of experiments can be conducted for other parameters also. The results of process validation must be maintained.

**6.4 Identification and traceability**

Organisation shall identify the product by suitable means throughout product realisation. Identify the product status with respect to monitoring and measurement requirement

if agreed contractually maintain traceability.

Material in production flow does not necessarily indicate identification.

Products at all stages of manufacturing must be suitably identified. The identification may be by,

- Colour codes
- Tags
- Marking
- Any other means

Identification is carried out in such a way that product of different status must be clearly recognized and shall not be mixed with each other.
Traceability must be maintained if it is a specified requirement of customer. The record of traceability must be maintained.

**Customer property**

Exercise control over customer property by receipt verification, identification and proper storage and protection. If customer property is lost or found to be unsuitable for use it shall be recorded and reported to customer. Customer property can include intellectual property.

**6.5 Preservation of product**

Organisation shall preserve the conformity of product during internal processing and delivery to the intended destination. The preservation shall include identification, handling, packaging, storage and protection; preservation shall also apply to the constituents parts of products

Products at all stages of manufacturing and up to destination must be preserved against adverse environmental conditions, mishandling, losses in transit and other unforeseen reasons. The preservation can be carried out by following means,

- Identifying product by suitable means

- Handling of the product shall be carried out in such a way that the product shall not get damaged, lost etc.wherever possible use of material handling equipments shall be made. Use of chutes and gravitational means is encouraged.
- Products shall be stored at safe and suitable area. Products stored shall be verified at defined interval to ensure condition of material.
- Packing shall be carried out as per customer's requirement. Packing instructions shall be prepared and utilized at the point of use giving details about type of material to be used box size, qty/ box etc.

# 7 CONTROL OF MONITORING AND MEASURING DEVICES

Determine measurement and monitoring to be undertaken and monitoring and measuring devises needed to provide evidence of conformity to product requirements.

Exercise control over monitoring and measuring devices by following means,

- Identification of measuring and monitoring device(MMD)
- Decide calibration frequency based on usage
- Prepare calibration schedule based on calibration frequency
- Calibration of MMD as per schedule from recognised lab.
- Safeguard against unauthorised adjustment
- Indicate calibration status
- Storage at proper place
- Initiate actions when MMD found out of calibration within due date.
- Maintain records of calibration.

Select measurement to be made and suitable instruments for measurement. While selecting measurements, consideration must be given to customer's requirements. It is recommended that measuring instruments must be ten times more accurate than the parameter, which is being measured.

Identify all the instruments by unique numbering system

e.g. ABC / OM / 01

Where ABC – Company abbreviation

OM – outside micrometer

01 -- Serial number.

Indicate these numbers on instruments by engraving, colour codes etc.

Specific instruments can be selected for specific application.

Means for measuring critical parameters, for final inspection, for raw material inspection

Dedicated instruments can be used and can be identified accordingly.

Decide calibration frequency of instruments based on following,

- Extent of use of the instruments
- Criticality of the parameter being inspected
- Environmental conditions during measurements

Prepare list of all measuring instruments including instrument name, number, least count, and calibration frequency etc.inlude all instruments used for measurement like templates, weighing scales, masters

After deciding calibration frequency prepare calibration schedule of instruments indicate calibration month and week of calibration in calibration schedule

Do calibration of instruments as per schedule.

While giving instruments to external lab for calibration ensure following,

- Identification of all instruments
- Calibration frequency
- Any specific requirements of calibration ex. For measuring pins calibration at particular points etc.

Calibration can be performed internally or from external accredited lab. While selecting calibration lab for instruments verify following,

- Infrastructure of the lab.
- Competency, experience of the personnel performing calibrations
- Accuracy of the instruments
- Environmental conditions maintained in the lab
- Accreditation obtained and its validity.
- Scope of the accreditation

External calibration lab must be accredited to ISO 17025/NABL or government approved lab or customer-approved lab. If for a particular instrument calibration facilities are not available then calibration can be performed by original equipment manufacturer

Indicate calibration status of the instruments by affixing sticker on instruments giving calibration date and due date. Charts can also be displayed at the point of use to indicate calibration status. Calibration status can also be identified by assigning colour codes to the instruments as follows
Fix colours codes for different category of instruments. Instruments having due in particular month can be identified by particular colour. I.e. instruments due in January by red colour, due in February yellow colour etc.

Verify certificates given by the calibration lab for following,

- Traceability of the calibrating instruments
- Total error observed
- Correctness of the other data

It is very essential to mention acceptance criteria for all instruments because if the instruments show error and if it is rejected as per standard, even then, you can use the instruments for measuring less accurate parameters but in this case, it is advisable to mention acceptance criteria.

Handling and storage of the instruments should be carried out in such a way that these shall not be damaged or accuracy is disturbed. Training to all concern must be given about handling and storage of the instruments. Measuring instruments should be stored,

- Away from dust, dirt, oil
- Away from heat and humidity and vibrations.
- In suitable boxes/ covers
- At identified and designated locations

Safeguard instruments against unauthorized adjustment which invalidate calibration setting by one or more of the following means

- Locking adjusting screws
- Restricting use to authorized personnel only

Many times, it happened that measuring instrument shows error within calibration date in this case initiate following actions as appropriate,

- Withdraw all items measured by affected instruments.
- Inform customer if such items are dispatched.
- Re inspect all items by fresh calibrated instruments
- Revise calibration frequency if required

### 7.1 Calibration records

Records of the calibration activity for all gauges, measuring and test equipment including employee and customer owned gauges shall include,

- Equipment identification including the measurement standard against which the equipment is calibrated.
- Revisions following engineering changes
- Any out of specification readings as received for calibration/verification
- An assessment of the impact of out of specification condition
- Statement of conformance to specification after calibration/verification activity
- Notification to customer if suspect product or material has been shipped.

Maintain records of the calibration in history cards of the instruments, which gives details about following,

- Instruments name and number
- Calibration agency and report number
- Calibration date and due date
- Actual values of calibration
- Acceptance criteria

# 8. SUPPLIER EVALUATION AND CONTROL

**8.1 Supplier evaluation and selection –**

**Procedure**

**Purpose and Scope**

To describe the process and method by which the company's suppliers (the term also includes contractors and subcontractors) are evaluated, selected and controlled.

Supplier Evaluation

Form **Procedure**

1. The selection criteria for suppliers is as follows:
   a. Ability and preparedness to meet IMS requirements.
   b. Ability and capability to meet legislative obligations and relevant industry and government standards and codes.
   c. Qualifications, experience and capability within the scope they are contracted for.
   d. Quality, consistency and reliability of product or service provided.
   e. Delivery performance.
   f. Price of product or service including commercial arrangements.
   g. Quality, environmental, health and safety management systems.
   h. Past performance including health, safety and environmental record.
2. Suppliers are classified on the basis of the potential risk their products or services may pose to:
   a. Products or services.
   b. Workplace health and safety.
   c. The environment.
3. Consideration needs to be given to whether:

a. Failure of the supplier to deliver agreed products or services will impact upon profitability?

b. Failure of the supplier to deliver agreed products or services will result in failure to meet any contractual, legislative and statutory obligations for delivery of products and services?

c. Supplier will introduce or potentially introduce any high risk hazards or significant environmental impacts to the workplace?

4. Suppliers that are identified as having the potential to significantly affect activities, products or services are deemed "critical" and must undergo a thorough documented evaluation and re-evaluation process. Non-critical suppliers are required to be evaluated but not necessarily to the same extent as those deemed to be critical. Examples of critical suppliers would include:

   a. Suppliers of Hosting Services.

   b. Suppliers of IT services.

5. The following rating system will be used for suppliers:

   a. Critical

   b. Approved

   c. Approved and inducted

   d. Back-up

6. Where applicable, Suppliers must have current and appropriate insurance arrangements in place. Certificates of currency for required insurances are to be provided as part of the formal evaluation process.

7. Products and services essential to meet contract requirements shall only be purchased from qualified and approved suppliers.

8. The details of suppliers are specified in the Suppliers data base,

inclusive of their rating. Supplier Induction

1. All Suppliers and their staff are to be effectively inducted, including training with respect to specific site procedure requirements. Refer to the Training, Competence and Awareness procedure for further details.

2. A record of the induction training conducted in to be retained in Supplier

Employee data base.

Re-Evaluation

1. Once evaluated and approved, suppliers are to be subjected to formal periodic re- evaluation. Re-evaluations are scheduled within the Supplier data base.
2. Re-evaluation is to take place at least once every two years or sooner if reasons apply. Some reasons for early re-evaluation are:
    a. Incidents and/or poor performance involving the supplier or contractor.
    b. Change in circumstances or structure such as new ownership, or change of location or key personnel.
    c. Change in scope of services.
3. Re-evaluation is to follow the same process as for the initial evaluation.

# 9. HEALTH AND SAFETY

## 9.1 Hazard Identification, Assessment and Control

**Purpose and Scope**

The purpose of this procedure is to effectively and systematically identify, document review and control actual and potential hazards onsite.

**Procedure**

1. Hazard Identification - On identification of a new hazard it will be entered into the Report a Risk data base along with the potential harm that the hazard is likely to cause.
2. Risk Analysis - The data base will step you through the process of assessing and developing appropriate controls based on significance.
3. Once the hazard has been entered and controls set, attach the new hazard to the appropriate register.
4. Once attached, print off the updated register and circulate to the appropriate employees or areas.
5. Register Review – management representative will notify Management when a register is due for review. The management will review any changes that have a bearing on the health and safety of employees to ensure that procedures can be updated or training arranged.
6. Monitoring of Hazards – Management representative will notify the management/concern person when any hazard monitoring is due. Examples of this include noise monitoring, atmospheric monitoring

Identification

1. Hazards may be identified by the following methods
    a. Initial and routine plant inspections of the workplace.

b. Employee may raise the hazard concern at a meeting
c. Hazard is drawn to the attention of the Safety rep, Supervisor or Manager.
d. Accident and incident or critical event identifies a new hazard.
e. New equipment and tasks will be subject to a hazard review before use

Assessment

1. Hazards will be assessed for significance/ risk.
2. Risk assessment matrix and risk scores will be recorded for all new hazards reported and will also be recorded in hazard registers.

Documentation and registers

1. Hazards that are significant are documented in the company hazard register
2. Task analysis (TA), Job safety analysis (JSA), Safe Work Statement Methods (SWMS) or Site Specific Safety Plans (SSSP) may also be used to identify hazards and document the associated controls.

### 9.2 Application of controls

Hazards that are significant will be controlled by applying all reasonably practicable steps. Controls will be applied in the following order

1. Elimination
2. Substitution
3. Engineering controls
4. Administrative controls
5. Personal protective equipment

Induction and training of staff in relation to hazards

1. All new staff will be inducted and sign an induction record
2. The hazard reporting process will be explained
3. The expected work methods, standards and controls will be covered at

induction

Induction of visitors, suppliers and sub-contractors

1. Visitors, suppliers and sub-contractors will typically sign in at reception and be inducted on site.
2. The induction content typically relates to the hazards they will be exposed to during their visit as a minimum general hazards and emergency procedures are covered.

Review of hazards onsite will completed as follows:

1. On a determined frequency (annual as a minimum) the hazard register is reviewed and updated if necessary. Review is documented in the Event Management Database
2. Work sites are reviewed and records maintained
3. Vehicles are checked and records maintained
4. Workshops and offices are checked and records maintained

Critical events

1. Critical events are recorded on an accident, incident or near miss form.
2. The investigation and actions are recorded and these events are tabled at the safety committee meeting for review.
3. The need for health monitoring may arise from a critical event and this monitoring will be completed and results supplied to the employee.

Purchasing policy and associated reviews of new equipment and processes

1. The company shall assess and record at pre purchase and upon arrival or commissioning stage the relevant H & S aspects of the plant, process or substances.
2. Refer to purchasing procedure. Specialist advisors and external

audits

3 Hazards that require specific specialist advice or monitoring then the Management Representative may authorise these services.

4.A list of specialist advisors is prepared and maintained

5. External audits may be commissioned as requested by the Managing Director

**9.3 Legal compliance and access to relevant information**

1. The compliance data base is used to record all legal compliance.
2. A review of compliance with relevant and new legislation or industry guidelines will be conducted annually.
3. Access to relevant information and codes of practice can be sources from the following:
    a. Documents folder
    b. Legislation
    c. Industry guidelines and approved codes
    d. Industry publications
    e. Other web sites
    f. Health monitoring Consultants

1. The control of hazards may involve the regular monitoring of employees or the workplace
   e.g. noise levels and audiometer. The frequency and type of monitoring (including exit testing) is detailed in the table below and typically established as an event
   Employees will be required to provide written and signed consent to the health monitoring provider.
2. The health monitoring provider may release relevant health information to employer (PCBU) so the employer can execute their duties under the Act and ensure that hazards are adequately controlled to protect workers from harm.
3. Results for monitoring - Results for monitoring in the workplace will be made available to employees.

4. Sub optimal or adverse results from monitoring will be reviewed and necessary controls applied and employee's medical and vocational needs will be

considered.

## 9.4 Personal Protective Equipment

1. The company will provide:
   a. All PPE required to protect employees from hazards while at work.
   b. The necessary PPE if worn out or expired and parts for basic maintenance.
   c. The necessary training in respect to the wearing or operation of the PPE
2. The PPE provided will comply with all relevant standards
3. PPE issued is recorded on the:
   a. Induction record
   b. Human Resources data base
4. An employee may provide PPE for their own use. This will be noted on the PPE issued record and signed. If the PPE supplied by the employee is not of the required standard or worn out the employee will be requested to replace it.
5. The hazard register or task analysis/ SWMS will detail the necessary tasks/ machinery hazards that require PPE to be worn.
6. The company will take all practicable steps to ensure employees wear PPE.
7. Employees failing to wear PPE as instructed may be subject to the companies' disciplinary procedure.
8. Visitors will be issued PPE necessary for the hazards and area they will be visiting.
9. Contractors and subcontractors will provide and wear PPE appropriate to the task they are completing.
10. A list of the standard issue of PPE is held in PPE module

**9.5 Safety Data Sheets (SDS)** previously known as (MSDS - Material Safety Data

Sheets)

1. The Management Representative will obtain Safety Data Sheets from the relevant approved suppliers.
2. Controls outlined in the SDS will be incorporated into training, documentation and site procedures where necessary.
3. Employees will be suitably trained.
4. MSDS will be filed in an appropriate documents folder.
5. Where requested by the Site Specific Safety Plan (SSSP), site owner or main contractor, copies will be maintained on sites for reference.

### 9.6 Accidents / Incidents reporting, recording and investigation

**Purpose and Scope**

Orrganisation will ensure an active reporting, recording and investigating all incidents and accidents.

**Procedure**

1. All accidents, incidents or near misses (involving injury, illness, persistent or unusual pain) will be reported early and promptly on the accident report form.
2. If necessary, the trained first aiders on site will attend to the accident.
3. The employee will forward all Medical documents to the Co-coordinator who will initiate action and file these as required.
4. All Notifiable events need to be advised to the Co-coordinator as soon as practical
5. The accident scene of a "Notifiable events" is not to be interfered with or disturbed until given a clearance.
6. An accident investigation may be required (refer to below)
7. Rehabilitation plans will be established as outlined in the rehab procedure.

Accident Investigation

1. Only people with the appropriate skills and experience should investigate

accidents.

2. Gather all the facts; all investigations will be initially recorded on an Accident / Incident register. for investigation pl ensure the following activities,
    a. Interview witnesses and describe events in detail, using any photos, diagrams or other exhibits that may be appropriate.
    b. Inform the prescribed agencies
    c. Be sure that you understand the sequence of events fully before any analysis takes place.
    d. Identify all the hazards involved.
    e. Consider:
        i. equipment,
        ii. materials,
        iii. work practices and procedures,
        iv. work environment,
        v. health issues,
        vi. hazards
3. Assess the Hazard Controls in place. What controls were in place, and why didn't they work? What is needed? Is there a need to train or inform employees?
4. Decide on Future Action. Describe fully what needs to be done to prevent further accidents or incidents. Who should do what and by when?
5. Records of those who hold relevant investigation skills are recorded in Human Resources data base
6. Inform all those affected. Inform everyone who needs to know, not only those directly involved. Health and Safety meeting agenda will include the results of any findings and actions to be undertaken.
7. Follow up. There must be checks to ensure that recommended changes have been made and results achieved.
8. The Manager or Supervisor will implement any corrective actions or

improvements that arise from the accident investigation. Actions must be signed off as they are completed.

## 9.7 Injury Management and Return to Work

1. Medical documents and medical certificates are received. They will be filed as necessary.
2. Employee having more than 5 days off and/or the injury would benefit from rehabilitation.
3. Signed consent form is required by the employee to authorise the release of information from the doctor.
4. Consult doctor or treatment provider as to possible duties the injured employee could undertake.
5. Employee, Employee's Manager/Supervisor and Co-ordinator meet and discusses possible options and formalise the options in a rehabilitation plan. The employee may have a Union or support person present.
6. A Third party rehabilitation provider may assist with the plan and monitoring of progress.
7. Rehabilitation plan and set milestones are monitored as agreed.
8. If rehabilitation milestones or the rehabilitation plan are not proceeding as agreed an occupational health specialist or physician may be involved. Typical transitional duties include:
    a. Admin/ paper work e.g. electrical compliance certificates.
    b. Non lifting site work
    c. Reduced hours per day
    d. Motor vehicle driving/ deliveries
9. Early intervention with a rehab programme is the best option. Remember that an employee has a fitness level and muscle toning from the type of work they have been doing and the longer they are out of the workforce the longer it will take them to get back to full fitness.

10. The threat of re-injury is also a serious possibility if the rehabilitation is not designed and handled correctly.
11. Remember an employee may feel threatened by the meeting format and the desire for the company to get them back to work earlier than they may be happy with.
12. Always offer the employee the option of having a support person or H & S rep present.
13. It is good policy to clearly explain to the employee that the companies aim is the smooth reintroduction back into the workforce.
14. Remember that high achievers and very active people can also overdo their return to work and reinjure themselves.

## 9.8 Employee Participation

**Purpose and Scope**

The organisation will ensure that employees have the opportunity to be fully involved in the development, implementation and operation of safe workplace practices.

The organisation actively encourages employees to be involved in the Health and Safety meetings.

**Procedure**

Meetings are scheduled in the Event Management data base. When completed, the event will be signed off and the minutes attached.

The following events may be used to facilitate participation:

Toolbox Meeting

1. The site foreman will chair and take minutes of site toolbox meetings.
2. The frequency of toolbox meetings is determined by the:
    a. Main Contractor/ Principal
    b. Site contract requirements
    c. Company directive

3. All employees and contractors on the site at the time of the meeting MUST attend unless excused by the foreman for an extraordinary reason.
4. Site foreman will record
    a. Names of all attendees
    b. Concerns or hazards raised
    c. Accident reported
    d. Concerns raised
    e. A brief summary of specific topics covered or instructions given
5. Completed site safety toolbox meeting records shall be held in Adhoc

Training register

. Health and Safety Meeting

1. The Health and Safety Meeting is made up of representatives from all levels within the organisation. The meeting can be a committee or it can be a full company meeting.
2. The meetings will be recorded with action points clearly identifying responsibility with target date for completion. The following items will be discussed:
    a. Previous minutes and actions taken
    b. Reviews of policies
    c. Correspondence, i.e. new laws and legislative requirements
    d. Objectives achieved
    e. Hazards/risk
    f. New equipment and new work processes (including hazards associated with new equipment or processes)
    g. Training undertaken and training for next period
    h. Accidents and incidents
    i. Upcoming and overdue events from Mango
    j. Changes that affect workplace safety
    k. General business Excellence

Appointment of Employee Health and Safety Representatives

If required annual nominations will be asked from employees for representatives to be elected. If more nominations are received for the positions available, an election will be held by ballot

Trained representatives have the following duties as outlined in the responsibilities section.

**9.9 Emergency Planning – Health and Safety**

**Purpose and Scope**

The purpose is to ensure that the organisation has an emergency plan to manage and test all types of potential emergencies.

**Procedure**

Emergency Plans

All potential emergencies will have document plans.

1. A copy of the emergency plan will be in a readily accessible location for all staff to refer to.
2. The emergency plan will incorporate all of the potential emergency situations that can affect the site that the plan has been developed for, and how to respond to them.

Testing Emergency Plans

1. Trial Emergencies will be scheduled in the Manage Events data base
2. When completed, the event will be signed off and the evidence attached.
3. All emergency wardens will undertake in house training every 12 months in their role and the emergency plan and procedures.
4. Training will be recorded on the individuals training record.
5. Any corrective actions or improvements will be recorded in the Improvement data base .Tests will be conducted every 6 months.
6. Following each test the Chief Warden will review the drill and implement any actions identified with the other wardens.

# 10 Environmental Aspects and Impacts Identification, Assessment and Control

## 10.1 Purpose and Scope

The purpose is to describe the procedures for identifying, assessing and controlling the environmental aspects and impacts.

**Procedure**

Aspects Identification

1. Ascertain the area(s) of the organization to be evaluated.
2. Identify environmental aspects via interviews, site reconnaissance, and document review for the area or function defined.
3. Consider the following;
   a. Waste streams (air, water, solid waste),
   b. Energy exchange (energy used and energy released),
   c. Resource consumption,
   d. Community
   e. Interested stakeholder input, and
   f. Human and Ecosystem toxicity.
4. Identify those aspects that are significant
5. Identify all resource consents.
6. Record the identified aspects in Risk

Management Data base Assessment and Control

1. The Risk Management data base will step you through the process of assessing and developing appropriate controls.
2. Once the aspect and impact has been entered and controls set, attach the new aspects to the appropriate register.
3. Once attached, print off the updated register and circulate to the appropriate employees or areas.

Review and Monitoring

EHS coordinator will notify the organisation when a review is due. This may be from a contractual requirement (for example: resource consent) or an operational site review.

## 10.2 Environmental Incident Reporting, Recording and Investigations

### Purpose and Scope

The purpose of this procedure is to ensure that environmental incidents are reported, recorded, investigated and appropriate corrective and preventive action is performed.

### Procedure

1. When an environmental incident is reported, an improvement will be raised in the Improvement data base
2. The workflow in the Improvement module, as setup then ensures the environmental incidents has corrective action assigned.
3. All actions will be recorded in the Improvement module.
4. When an environmental incident is reported, an improvement will be raised in the register. Examples of environmental incidents include:
    a. Discharge to air
    b. Discharge to land
    c. Discharge to water
5. Contain the environmental incident and prevent or minimise the risk of further environmental harm. Contact the appropriate Environmental Protection Agency.
6. The environmental incidents have corrective action assigned.
7. All actions will be recorded
8. The preventive actions are verified in the Management review meeting to ensure they are effective

## 10.3 Environmental Emergency Planning

### Purpose and Scope

The purpose is to ensure that the organisation has an emergency plan to manage and test all types of potential emergencies.

### Procedure

1. All emergency situations will have an Environmental Emergency Plan
2. A copy of the environmental emergency plans will be in a readily accessible location for all staff to refer to.
3. The emergency plan will incorporate all of the potential emergency situations that can affect the site that the plan has been developed for, and how to respond to them.

Testing Emergency Plans

1. Trial Emergencies will be scheduled in the Events register.
2. When completed, the event will be signed off and the evidence attached.
3. All emergency wardens will undertake in house training every 12 months in their role and the emergency plan and procedures.
4. Training will be recorded on the individuals training record.
5. Any corrective actions or improvements will be recorded in the Improvement record.
6. Tests will be conducted every 6 months.
7. Following each test the Chief Warden will review the drill and implement any actions identified with the other wardens.

# 11. PERFORMANCE EVALUATION

## 11.1 Monitoring, Measurement and Evaluation

### Purpose and Scope

To describe how we will monitor, measure, analyse and evaluate the IMS in order to identify and take suitable action to ensure the continual improvement of the management system.

**Procedure**

1. The company will determine:
    a. The aspects of the IMS that will be monitored and measured.
    b. The responsibilities, frequency and methods for monitoring, measurement, analysis and evaluation needed.
    c. The criteria against which we will evaluate its IMS performance.
    d. When the monitoring and measuring:
        i. Will be performed.
        ii. Results will be analysed and evaluated.
2. The results of the analysis and evaluation conducted is to evaluate the:
    a. Degree of customer satisfaction.
    b. Conformity of products and services.
    c. Performance and effectiveness of the IMS including the environment, health and safety and quality.
    d. If planning has been effectively implemented.
    e. Effectiveness of actions taken to address risks and opportunities.
    f. Performance of external providers.

g. Need for improvements to the IMS.

3. Appropriate documented information must be retained as evidence of the monitoring, measurement, analysis and evaluation that is conducted.

Monitoring Arrangements

1. Generally, individual procedures within the IMS describe the specific monitoring, measurement, analysis and evaluation requirements to be met.
2. Whenever required, an event will manage the process.

## 11.2 Internal Audit

**Purpose and Scope**

To describe the responsibilities and methods used to evaluate the effectiveness of the implementation and maintenance of the IMS.

Audits and inspections are completed for the following purposes:

- To identify the compliance status against the company policies, procedures, legal requirements and other obligations including the ISO 9001,
- To identify areas where IMS performance needs to be improved or changed.
- To identify leading practice, so as such practices can be communicated and implemented in other the company activities.

**Procedure**

The company will use the DIME matrix methodology for internal audits. This will be an audit of the system based on the ISO clauses.

The company will audit using the Compliance Data base to capture the records of DIME (documented, implemented. monitored, evidence/effective).

1. An audit of the IMS will be conducted as per the schedule in Event Management Module.
2. An event will remind Management when the audit is due.
3. The Audit team will audit the assigned section of the IMS.
4. An auditor can't audit a core function they are responsible for.
5. Once complete, the event will be signed off and a copy of the report uploaded.
6. An improvement will be raised for each non-conformance identified.

### 11.3 Management Review

**Purpose and Scope**

To ensure that the IMS is effectively reviewed on a regular basis with the purpose to:

- Ensure its continuing suitability, adequacy, effectiveness and alignment with the strategic direction.
- Consider relevant changes in external and internal issues, including changes to legislative and other requirements.
- Ensure all employees are aware of the current status of IMS performance and changes to the IMS system and procedures.
- Allow for all employees to provide suggestions, direction and resources for the continual improvement of IMS performance.

**Procedure**

1. A formal review of the IMS is to be conducted every six weeks, all staff

are invited and expected to attend in person or via appropriate communication.

2. The agenda will be:
    a. the status of actions from previous management reviews;
    b. changes in external and internal issues that are relevant to the quality management system;
    c. information on the performance and effectiveness of the quality management system, including trends in:
        i. customer satisfaction and feedback from relevant interested parties;
        ii. the extent to which quality objectives have been met;
        iii. process performance and conformity of products and services;
        iv. nonconformities and corrective actions;
        v. monitoring and measurement results;
        vi. audit results;
        vii. the performance of external providers;
    d. the adequacy of resources;
    e. the effectiveness of actions taken to address risks and opportunities
    f. Opportunities for improvement.
3. An event has been setup to ensure the Management Review happens.
4. The Management Representative will run, keep minutes and publish records in Event Management data base.
5. The Management review will follow the standard agenda format in the minutes.
6. Actions are assigned and recorded on the Management Minutes with agreed timeframes.

# 12. IMPROVEMENT

## 12.1 Improvement and CorrectiveActions

**Purpose and Scope**

To ensure that improvements, non-conformities and corrective actions are reported, recorded, investigated and followed-up.

The procedure also ensures that non-conforming products or services are identified, reported, recorded, investigated and controlled.

**Procedure**

1. Employees must report improvement opportunities, non-conformances, failures and any other IMS issues.
2. Improvements can be initiated by any employee when any of the following issues are identified:
   a. To initiate a change to the IMS.
   b. To initiate an improvement to the performance and effectiveness of the IMS.
   c. When an innovation or improvement opportunity is identified.
   d. When a non-conformance is identified at any time,
   e. When a discrepancy, non-conformance or improvement is identified duringauditing.
   f. When a customer complaint or any significant customer feedback is received (including compliments).
3. Improvements are to be recorded and retained including associated

documents and records with respect to the improvement

4. The improvement workflow will manage the improvement process
5. Findings will be reported to the Management Review meeting including their status.

SECTION 2

# PRECAUTIONS DURING CORONAVIRUS DISEASE PANDEMIC

This section includes practical recommendations to organizations and workers on how to manage risks arising out of pandemic and is suitable for organizations resuming operations, those that have been operational throughout the pandemic, and those that are starting operations.

The guidance is generic and applicable to organizations regardless of the nature of business, service provision, size or complexity. It recognizes that many smaller organizations do not have dedicated departments for functions such as occupational health and safety facilities management or human resources. The purpose of this document is help organisations to,

- take effective action to protect workers and other relevant interested parties from the risks related to CORONAVIRUS DISEASE;
- Demonstrate that it is addressing risks related to CORONAVIRUS DISEASE using a systematic approach;
- Put in place a framework to enable effective and timely adaptation to the changing situation.

The guidelines can be used along with the guidelines given in ISO 9001-2015 AND ISO 45000-2018

THE guidelines are helpful for the following type of organisations,

Those that have been operating throughout the pandemic;

- Those are resuming or planning to resume operations following full or partial closure;

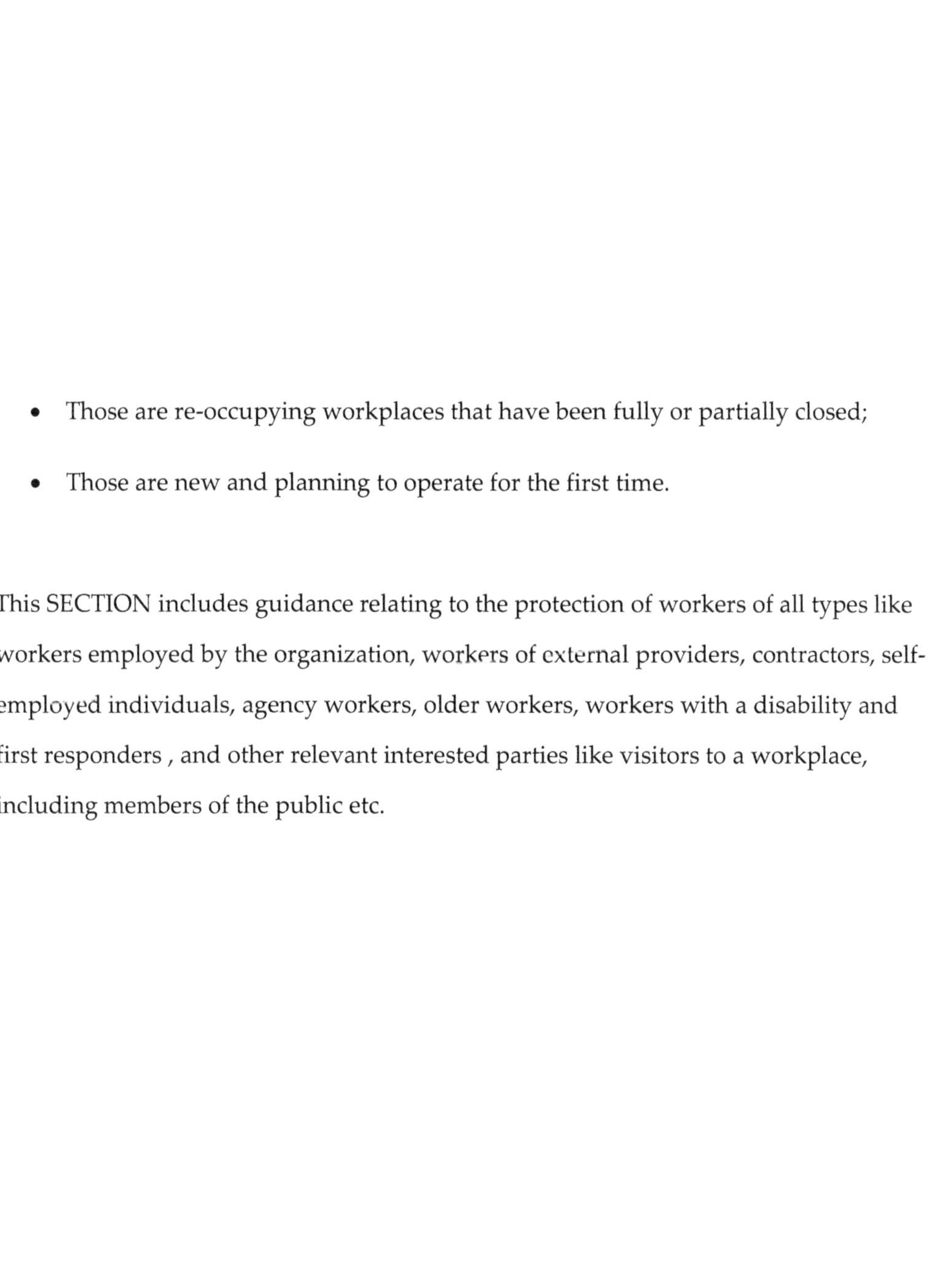

- Those are re-occupying workplaces that have been fully or partially closed;
- Those are new and planning to operate for the first time.

This SECTION includes guidance relating to the protection of workers of all types like workers employed by the organization, workers of external providers, contractors, self-employed individuals, agency workers, older workers, workers with a disability and first responders , and other relevant interested parties like visitors to a workplace, including members of the public etc.

# 1. RISKS TO WORKERS AND OTHER PARTIES

## 1.1 General

Consider the following parameters while understanding risks to workers and other parties

a) What can affect the ability of individuals to work safely during the CORONAVIRUS DISEASE PANDEMIC?

b) How its operations should change to address the increased risk to work-related health, safety and well-being.

Consider the following specific external and internal issues that can affect the health and safety of workers and how these issues are impacted by the pandemic.

External issues can include, but are not limited to:

- The occurrence of CORONAVIRUS DISEASE within the local community (including in other organizations and other workplaces);
- Local, regional, national and international circumstances, and related legal requirements and guidance;
- The availability of clinical services, testing, treatments and vaccines;
- The availability of health and safety and other supplies (e.g. PPE, masks, hand sanitizer, thermometers, cleaning and disinfection materials);
- How workers travel to and from work (e.g. public transport, car, bicycle, walking);
- Workers' access to childcare and schooling for their children;
- The suitability of a worker's home for remote working;

- Workers' domestic situations (e.g. living with someone who is considered to be at higher risk of contracting CORONAVIRUS DISEASE or getting severe illness from it,
- Changes or problems in the supply chain;
- The continuity of essential services like food provision, domestic infrastructure, utilities etc.
- Changes in customer needs and expectations, or behaviors;
- Local culture and cultural behaviors like Kissing, hugging, shaking hands etc.
- Increased or decreased demand for products/services.
- Internal issues can include, but are not limited to:
- The occurrence of CORONAVIRUS DISEASE in the organization;
- The number and types of workplaces (e.g. offices, factories, workshops, warehouses, vehicles, retail outlets, workers' own homes, other people's homes);
- Cultural values within the organization that can affect risk control measures;
- The ability of the organization to gain up-to-date knowledge about COVID-19;
- The type of organization and related activities (e.g. manufacturing, services, retail, social care, training or other education, delivery, distribution);
- The type of workers in the organization (e.g. employed, contractors, volunteers, freelance, part-time, shift workers, remote workers);
- The extent to which it is possible to implement physical distancing measures;

- Specific needs of workers (e.g. workers considered to be at higher risk of contracting CORONAVIRUS DISEASE or getting severe illness from the disease.
- Workers with caring responsibilities, disabled workers, pregnant women, new mothers and older workers;
- Increased worker absence due to sickness, self-isolation or quarantine requirements, bereavement;
- Resource availability, including adequate provision of toilet and hand washing facilities;
- How work is organized (e.g. changed work demands, pace of work, time pressure, shift work) and supported, and how this impacts work-related health, safety and well-being.

**1.2 Risk management and worker participation**

To assist effective management of the risks arising from CORONAVIRUS DISEASE relating to work, the organization should:

- Demonstrate leadership and commitment to collective responsibility and safe working practices;
- Communicate about, and consistently comply with, internal policy at all times;
- Commit to transparency when reporting and managing suspected and confirmed cases of COVID-19, ensuring that personal health information is kept confidential

- d) Make sure that adequate resources are provided and make them available to workers in a timely and effective manner;
- Make sure that consultation and encourage participation of workers and worker representatives, where they exist, in making decisions that affect work-related health, safety and well-being;
- Provide a clear policy on the financial implications for workers unable to work due to operational restrictions, or who are required to self-isolate or quarantine;
- provide appropriate support for workers unable to work due to operational restrictions, or who are required to self-isolate or quarantine, including provision of appropriate leave from work and paid sick pay if possible (so that workers do not come to a workplace when they should not because of concerns about pay);
- Communicate how workers and other relevant interested parties should report incidents or raise concerns, and how these will be addressed and responses communicated;
- protect workers from reprisals when reporting potential illness or incidents, or if workers remove themselves from work situations which they believe to be unhealthy or unsafe;
- Make sure the coordination across all parts of the organization when implementing measures to manage the risks related to COVID-19;
- Seek competent advice and information on managing risks related to COVID-19, if necessary.

### 1.3 Care to workers and other interested parties

The organization has a duty of care to workers and other interested parties who can be affected by their activities, including customers, service users and the general public. By encouraging wide input, the organization can have a better overview of risks to work-related health, safety and well-being during the pandemic. Active and ongoing engagement with workers and worker representatives, where they exist, is likely to result in better outcomes when managing the risks related to CORONA VIRUS DISEASE.

The organization should:

- Encourage participation and involve workers and worker representatives, where they exist, in assessing risks related to CORONAVIRUS DISEASE and making decisions on how to manage them;

- Communicate to workers and other relevant interested parties (e.g. the public, customers, suppliers, visitors, students, investors, shareholders, regulators, unions) how the organization is managing risks from CORONAVIRUS DISEASE(communication can be through any appropriate method, see Clause 9);

- Provide one or more ways for workers and other interested parties to give feedback on actions taken to manage work-related health, safety and well-being (e.g. through virtual meetings, collaboration tools, online surveys, emails);

- Take timely and appropriate action to address concerns raised by workers and other interested parties and communicate these actions to them.

The organization should make sure that decision-makers and worker representatives, where they exist, take into account the full diversity of the workforce and the specific experiences, views and needs of, for example, workers with disabilities, women, workers from different ethnic and faith groups, and workers of different ages.

**1.4** Planning **to eliminate risks**

Planning enables the organization to identify and prioritize risks arising from the pandemic that can affect work-related health, safety and well-being.

Although it is not possible to eliminate the risks related to CORONAVIRUS DISEASE entirely, planning should identify and prioritize the risks to workers in order to reduce those risks.

When planning, the organization should consider the following issues

- Practical changes that should be made to how work is organized and where work takes place;
- Interaction between workers;
- Interaction between workers and other people, including visitors, customers and members of the public;
- How to maintain complete and accurate contact information on people who interact closely like Workers in shifts, customers in pubs and restaurants, clients in gyms etc. for the purpose of contact tracing, respecting the need for confidentiality;
- the safe use of common areas and shared equipment;
- The impact of the pandemic on psychological health and well-being.

The organization should take a systematic approach to determining and addressing risks related to CORONAVIRUS DISEASE and identify work activities that:

- Can be done from home;
- Cannot be done from home, but can comply with physical distancing guidelines in the workplace, if practical adjustments are made;

- Cannot be done from home and cannot comply with physical distancing guidelines in the workplace.

For many organizations, the best way to mitigate work-related risks from CORONAVIRUS DISEASE is to enable and support workers to work from home, including in organizations that have fully implemented controls to protect against transmission of the disease. The organization should minimize the number of workers in a physical workplace, where this is possible, to provide enhanced protection through reduced contact with other people. The organization should take into account the needs of service users, clients and customers, as well as the workers performing the work, when determining the numbers of workers in a physical workplace.

The organization should make sure that additional support measures are implemented to protect the physical and psychological health and the well-being of workers who are working from home. The organization should consider if it is possible to enable a safe return to the physical workplace for individual workers if the home is not suitable, or if home working has a significant negative impact on their psychological health and well-being.

Work activities that cannot be done from home and cannot comply with physical distancing guidelines should only take place if the activities are essential and additional controls are implemented to mitigate the risks.

When planning to address risks related to COVID-19, the organization should take into account existing OH&S risks and measures already in place to manage these. The organization should:

- Assess if existing OH&S measures and controls need to be adjusted, taking into account any changes to work processes;

- Consider new OH&S risks like impact on fire safety arrangements) and other risks that can be introduced by implementing additional safety measures to manage the risks related to CORONAVIRUS DISEASE
- Plan actions to address new risks;
- Plan for changes in restrictions at short notice, whether at local, regional, national or international level, to minimize operational disruption

# 2. WORKPLACES, WORKING CONDITIONS AND ENVIRONMENT

## 2.1 Physical workplaces

The organization should make sure that workplaces (including all premises, sites and other locations where work takes place, including outside of a building) and facilities within those workplaces are clean and safe to use.

To prepare for safe operation, the organization should, as a minimum:

- Assess all premises, sites or parts of sites, including those that have been closed or partially operating;
- Establish arrangements to prevent potentially infectious people from entering the workplace display work instruction about the same at the entrance.
- Perform maintenance checks and activities on equipment and systems;
- assess and control risks related to air and other water-related diseases, in order not to introduce other health risks, particularly if water-based systems (including some types of air conditioning) have not been used for a period of time or if use has been reduced;
- Establish enhanced and/or more frequent cleaning and disinfection schedules, (e.g. by increasing the working hours and/or numbers of workers in cleaning roles, and encouraging other workers to clean and disinfect their own work zones and equipment regularly);
- provide enhanced personal hygiene facilities, including additional hand washing stations where possible and hand sanitizer points where this is not possible (including outdoor areas used for work or breaks), ensuring these facilities are accessible to workers with disabilities;

- coordinate and cooperate with other organizations on shared sites, including with contractors, managing agents, landlords and other tenants, ensuring both routine operations and emergency plans are taken into account.
- Deep cleaning and disinfection of workplaces and equipment;
- Disinfecting taps, showers and other sources of water with products that meet official requirements for use against COVID-19, and flush through before use;
- Maximizing the amount of outdoor air and room air changes through ventilation systems (with appropriate filtration and duration of operation), turning off air recirculation systems, and keeping doors and windows open to the extent possible;
- Ensuring toilet facilities are managed to facilitate safe use
- Restarting and testing specialist equipment that has been unused for longer than usual;
- Testing fire safety systems, including battery-powered units such as emergency lighting and alarms;
- putting in place signs and floor and/or wall markings to indicate recommended physical distancing, ensuring markings are simple, clear and large enough to be seen by visually impaired people;
- putting in place physical barriers to enforce physical distancing to the extent possible, where it is safe to do so without introducing new OH&S or other risks or negatively impacting people with disabilities;
- Creating work zones to limit the number of people in any one area (see 12.5);

- Limiting the number of people using shared equipment by creating working teams or pairs and assigning them to designated shared equipment;
- Establishing cleaning and disinfection points to enable workers to wipe surfaces and equipment throughout working hours;
- Reorganizing moveable equipment, desks and workstations to enable physical distancing;
- Fixing doors open to reduce touching of door handles (excluding doors required for fire safety, security or privacy);
- Establishing processes for safe entry and exit from workplaces;
- establishing one-way systems in corridors, stairways and other common areas, putting in place signs and floor or wall markings, and taking other actions to mitigate the risks where this is not possible;
- Determining safe ways of using lifts/elevators, including limiting capacity, and ensuring guidance for safe use is communicated both inside and outside of lifts/elevators;
- Providing additional outside spaces for workers to use for routine work, meetings and breaks, where possible.

**2.2 Working from home**

The organization should enable workers to work from their own home wherever possible, as this is one of the most effective ways of managing the risks related to the pandemic. The organization has the same responsibility for the health and safety of workers who are working from home as it does for those in a fixed physical workplace.

The organization should take all practical steps to remove barriers to working from home.

In determining which workers should work from home, the organization should ensure that whether worker;

- Can effectively perform role from home?
- Home situation suitable for home working?
- Want to return to a physical workplace?
- Can travel safely to and from a physical workplace without significant exposure to CORONA VIRUS DISEASE

The organization should consult with the worker to systematically assess the risks related to working from home and the actions needed to address the risks, as far as practicable, taking into account factors such as:

- the domestic circumstances of the worker like childcare or other caring responsibilities, domestic abuse, household members considered to be at higher risk of contracting CORONAVIRUS DISEASE or getting severe illness from it,
- The physical suitability of the home such as size, other people sharing the space, noise levels, suitable lighting, ergonomic issues etc.
- If the worker has access to relevant systems and information like. email, shared electronic drives, databases, enhanced security on relevant systems and guidance on operating securely while at home;
- The need for ongoing support for the use of IT equipment and software such as online conference tools

- The potential need to allow workers to take equipment that they use at work home on a temporary basis or to provide additional equipment like computer, computer monitor, keyboard, mouse, ergonomically suitable chair, footrest, lamp, printer, head set etc;
- The need for guidance on setting up an ergonomically suitable home workstation like enabling good posture and encouraging frequent movement;
- Psychosocial risks
- Impacts on personal or home insurance and tax liabilities.

The organization should provide workers with guidelines on what to do if the worker or any member of the worker's household is exposed to or contracts CORONAVIRUS DISEASE and is required to self-isolate or quarantine.

**2.3 Working in other people's homes**

Workers should not perform work activities in other people's homes if someone in that household has symptoms of CORONAVIRUS DISEASE(or is self-isolating or in quarantine) or is considered to be at higher risk of contracting CORONAVIRUS DISEASE or getting severe illness from it, except:

- To provide essential health and personal care such as medical or social care workers etc.
- To remedy a direct risk to safety or security such as emergency repairs by a plumber, construction worker, electrician, gas engineer etc.
- To address an issue in the home where this can be performed with additional social distancing or other measures to protect the vulnerable person.

- When preparing for workers to perform activities in other people's homes, the organization should:
- check if anyone in the household has symptoms of COVID-19, is self-isolating or in quarantine, or has been advised to isolate from other people to protect themselves because they are considered to be at higher risk from COVID-19;
- Consider if the work can be performed using digital or remote alternatives like video or phone consultations;
- communicate with households prior to work commencing, to discuss and agree how work will be carried out and general practices to minimize risk (e.g. how to enter the building without face-to-face contact, sanitizing hands before entering the household and washing hands before exiting, maintaining physical distancing while in the home, leaving internal doors open to minimize contact with door handles);
- Assign workers to work in household's local to them, wherever possible, to minimize travel and use of public transport;
- make sure workers have access to adequate PPE, masks or face coverings, hand sanitizer, cleaning and disinfection materials;
- allocate the same individual, pair or small team of workers to a household if repeat visits are necessary or the work is ongoing taking into account the type of work activities and the amount of contact those workers have with other people outside of the household.

The organization should establish and communicate a clear policy and process to manage situations where workers are required to self-isolate or quarantine due to one or more individuals contracting CORONAVIRUS DISEASE or being exposed to someone with CORONAVIRUS DISEASE

Working in multiple locations or mobile workplaces

The organization should make sure that workers with roles that cannot be performed at home or in a fixed physical workplace like drivers, social and personal care providers, cleaners, postal workers, delivery workers, traffic wardens, repair and maintenance workers are given support, guidance and adequate resources to work safely and to avoid transmission of the disease through travel and interaction with other people.

The organization should consult with workers and worker representatives, where they exist, to make sure that workers with mobile roles are fully informed and confident to use their own discretion to act appropriately in different situations. The organization should provide guidance and encourage workers in mobile roles to:

- Follow the guidance on physical distancing and hygiene
- Follow guidance on how to act in situations where physical distance cannot be maintained, or is not maintained by other people;
- follow guidance on how to act if other organizations require the removal of PPE, masks or face coverings for security or other reasons;
- make sure they have access to sufficient PPE, masks, face coverings, hand sanitizer, cleaning and disinfection materials, as appropriate;
- Follow guidance on how to access and safely use resources such as public toilets, and how to safely procure and consume food and drink;
- Retain documented information to support contact tracing, if necessary, about the places they go to in the course of their work;
- retain details of the people they have prolonged interaction or close contact with, where possible, to support effective contact tracing if a worker or other relevant

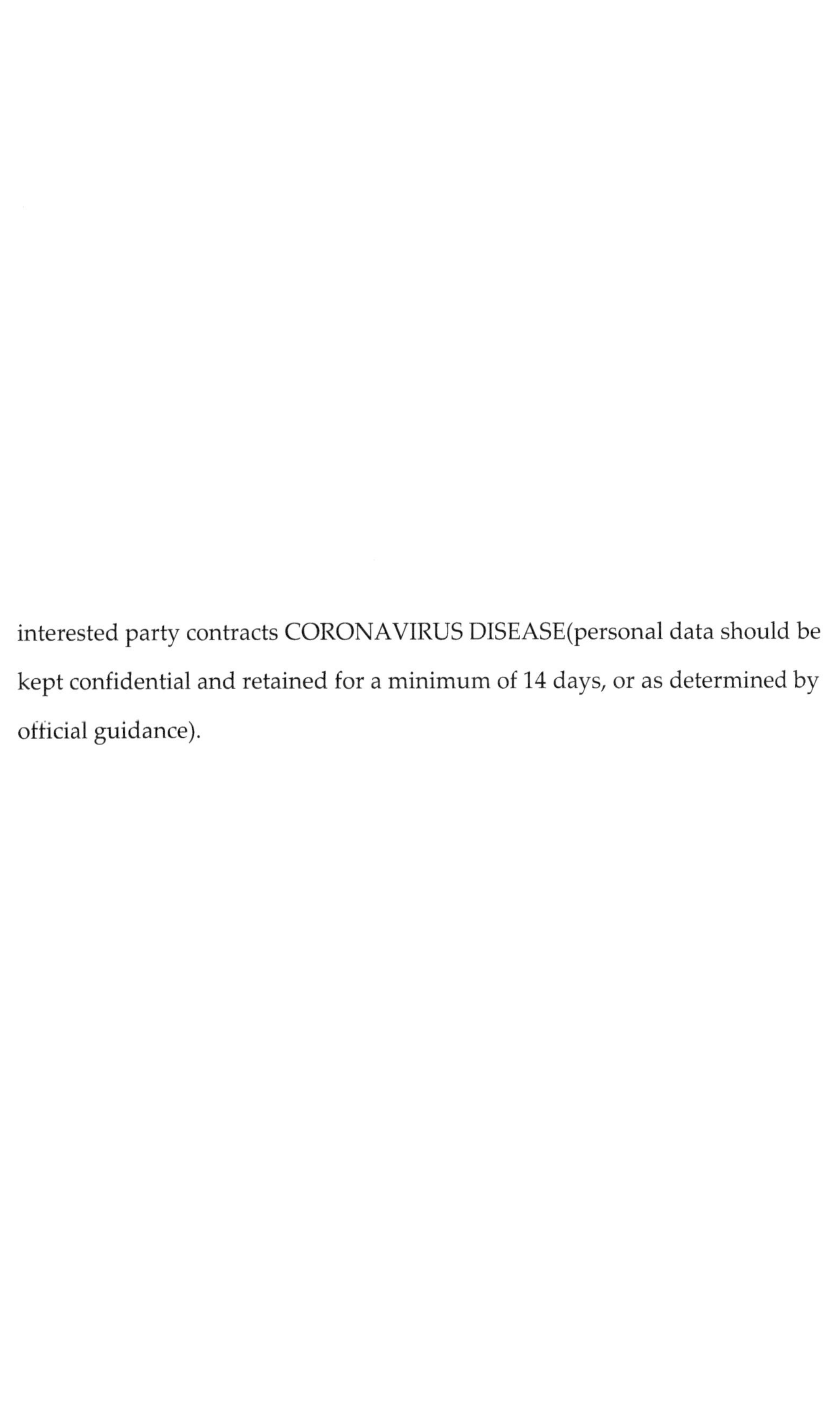

interested party contracts CORONAVIRUS DISEASE(personal data should be kept confidential and retained for a minimum of 14 days, or as determined by official guidance).

# 3. WORKERS ROLES AND ACTIVITIES

### 3.1 Assessing roles

In assessing roles, activities and where a worker should work, the organization should take into account workers who:

- Are considered to be at higher risk of contracting CORONAVIRUS DISEASE or getting severe illness from it;
- Are caring for someone who is considered to be at higher risk of contracting CORONAVIRUS DISEASE or getting severe illness from it;
- Are in a household with someone who is considered to be at higher risk of contracting CORONAVIRUS DISEASE or getting severe illness from it;
- are entitled to, request or need additional reasonable adjustments due to disability or other individual circumstances like conditions such as autism, pregnancy, disproportionately affected minority groups etc;
- Need additional support to protect their psychological health and well-being.

The organization should support workers with roles that can be performed effectively at home to work from home. To make sure this is effective, the organization should take actions determined by the consideration of issues and establish regular virtual or phone meetings to provide support, monitor well-being and make sure workers are connected to other workers, including those working on-site. The organization should make sure that there is clarity about what is and what is not expected of workers working at home and accommodate individual worker needs as far as possible.

For workers who need to be in a physical workplace, the organization should:

- Determine which roles are critical for operational continuity, safe facility management or regulatory requirements and cannot be performed from home;
- Identify workers in critical roles who are unable to work from home due to home circumstances or the unavailability of specialist equipment;
- Determine the minimum number of workers needed in a physical workplace at any one time to operate safely and effectively;
- Determine how activities are organized such as reducing job rotation, requiring workers to perform one activity with one set of equipment throughout the shift, enabling flexible working hours etc.

The organization should offer workers considered to be at higher risk of contracting CORONAVIRUS DISEASE or getting severe illness from it, and who cannot work from home, the option of the safest available roles in the physical workplace. Such roles should allow workers to maintain physical distancing guidelines at all times. If workers considered being at higher risk of contracting CORONAVIRUS DISEASE or getting severe illness from CORONAVIRUS DISEASE cannot comply with physical distancing guidelines, the organization should consult with the worker and worker representatives, where they exist, to assess if there is an acceptable level of risk if additional safety measures and controls are implemented.

The organization should consider assigning specific workers (or a single worker, in a small organization) the responsibility for ensuring CORONAVIRUS DISEASE safety measures and controls are implemented and maintained and for reporting issues to top management.

If workers are allocated new roles or tasks, the organization should provide adequate training and support to make sure that workers are competent to perform those roles.

The organization should monitor the introduction of safety measures or controls for any unjustifiable negative impact on some groups compared with others (e.g. workers with caring responsibilities, workers with religious commitments, workers with disabilities, pregnant workers).

**3.2 Activities to reduce risk of transmission**

If physical distancing guidelines cannot be complied with for a critical activity, the organization should take all possible further mitigating actions to reduce the risk of transmission of CORONAVIRUS DISEASE between workers and through interaction with other people in the workplace.

Before resuming work, the organization should take mitigating actions, such as:

- establishing fixed small teams or pairs of workers to limit the number of people in close contact; teams or pairs should be treated as a unit if any worker develops CORONAVIRUS DISEASE symptoms and all members of the unit should self-isolate or quarantine according to official guidance;
- revising work instructions to enable safe operation of activities like keeping activity times as short as possible, using screens or barriers to separate people, using back-to-back or side-to-side working instead of face-to-face etc;
- Establishing distinct zones for work activities that cannot comply with physical distancing guidelines;
- Using isolated spaces to enable physical distancing for workers who can safely work alone;
- identifying activities where workers directly pass objects like job information, spare parts, samples, purchased items; to each other or to other people, including the public, and establish processes to remove direct contact if possible.

- Providing appropriate PPE and guidance on how it should be used.

### 3.3 Emergency preparedness and response

- The organization should prepare for foreseeable emergencies and assess and revise existing processes as necessary.
- The organization should consider, for example:
- emergency processes like guidance on evacuating in teams to limit close contact with others, adjusting how workers and other relevant interested parties are required to assemble to increase physical distancing between teams etc;
- Reviewing personal emergency evacuation plans for people with assisted or facilitated evacuation needs including provision of additional PPE as necessary
- Training additional people to respond in emergencies, in case illness, self-isolation or quarantine results in a shortage of trained workers in the workplace;
- Providing first aiders with personal first aid resources, including appropriate PPE, in case of medical emergency or accidents;
- Providing clear guidance on processes for dealing with aggressive or violent people.

In an emergency where there is immediate danger (e.g. chemical spill, fire, break-in), complying with physical distancing guidelines can be challenging. Immediate preservation of life should be prioritized; however, the organization should also amend emergency plans to mitigate the risk of transmission of CORONAVIRUS DISEASE in emergency situations, as far as reasonably practicable.

The organization should assess additional risks that can arise from challenges to physical distancing during fire drills, simulations or other practice exercises and raise awareness of amended emergency plans. When planning for these exercises, the organization should make sure that additional safety controls and measures are in place if physical distancing guidelines cannot be maintained during, for example, evacuation from the workplace.

The organization should require workers who provide assistance to others in emergency situations to take additional and immediate hygiene measures following the emergency event, including hand washing or sanitizing.

**3.4 Monitoring and Planning of risks**

The organization should make sure that current and emerging risks related to CORONAVIRUS DISEASE are monitored and plan for occasions when restrictions are likely to be changed at short notice

The organization should determine actions it can take to enable a rapid and effective response to changes in restrictions to continue operations as far as possible. Planning should take into account different potential situations, including increased or different restrictions, or the lifting of restrictions. Planning should be undertaken in consultation with workers and worker representatives, where they exist

When planning, the organization should consider:

- Reducing operations to core activities that can be carried out with full physical distancing by a minimum number of workers in the physical workplace or by home-based workers;

- Whether operations can be modified to enable the organization to continue to work during periods of restriction;

- Whether full or partial suspension of operations is needed to consider the correct actions to take (e.g. pause operations to put in place additional measures or to reorganize work activities);
- Whether alternative operations can be implemented;
- The potential impacts on workers, taking into account workers with specific needs and circumstances;
- How individual workers can be impacted by different location restrictions such as workers who need to cross local, regional, national or international boundaries;
- The potential impacts on the supply chain and actions necessary to manage these;
- The need for cooperation and communication with partner organizations, organizations sharing facilities and other relevant interested parties.

The plans for different types of restrictions should address how to:

- Agree and communicate which workers:
    1. Will be required to be on-site;
    2. Will be required to work from home;
    3. Will not be able to work at all;
- Communicate the likely impact on working hours, pay and other conditions;

- Communicate to customers and other interested parties how changes to restrictions will affect operations through social media, apps, signage, websites etc.

The organization should take into account the individual impact on workers who are unlikely to be able to work at all if certain restrictions are imposed and inform them of the possible or likely impact on pay or employment conditions.

The impact on workers of a sudden easing of restrictions should also be taken into account (e.g. ability to return to work at short notice due to childcare responsibilities, workers considered to be at higher risk of contracting CORONAVIRUS DISEASE or getting severe illness from it, or living in households with higher risk people, workers self-isolating or under quarantine at that time).

Plans should be communicated to workers and other relevant interested parties at the earliest opportunity.

# 4. MANAGING SUSPECTED OR CONFIRMED CASES OF CORONA VIRUS DISEASE

## 4.1General

The organization should establish and communicate processes to manage suspected and confirmed cases of CORONAVIRUS DISEASE

To limit possible introduction of CORONAVIRUS DISEASE into the workplace, the organization should implement measures to assess people entering the building and prevent entry by those who have symptoms, who have recently travelled to or from areas with significant community spread of the disease, or who have been exposed to individuals infected with CORONAVIRUS DISEASE.

Top management and managers at all levels should support workers to take immediate action to self-isolate if they develop symptoms of disease, or quarantine if required to do so, and understand the processes in place and what is expected of them in relation to reporting, self-isolation or quarantine, and return to work.

Outbreaks of CORONAVIRUS DISEASE in the organization should be notified to relevant regulators and health authorities

**Managing illness in a physical workplace**

To minimize transmission of CORONAVIRUS DISEASE and to protect first responders, including first aiders, and the person they are treating, any person who becomes unwell in the workplace should be treated as a potential CORONAVIRUS DISEASE case.

The organization should consult workers with first aid responsibilities to determine if they are willing and able to continue to perform this role, taking into account individual circumstances .if the worker is considered to be at higher risk of contracting CORONAVIRUS DISEASE or getting severe illness from it, is living in a household with someone at higher risk, or if the worker has anxiety about increased exposure.

The organization should:

- Provide suitable PPE like face shields, gloves, gowns and masks and give guidance on how these should be used by first aiders
- Isolate the person who is unwell while first aid is provided or if transport from the workplace needs to be arranged (e.g. transport can be provided by a member of the same household);
- Provide the affected person with a mask (consideration should be given to people with underlying health conditions that affect breathing) and ask them to wash or sanitize their hands;
- Require the affected person to leave the workplace, using a safe method of transport, avoiding public transport if possible, to a suitable safe place.

- advise the affected person to request a CORONAVIRUS DISEASE test if they have recognized symptoms and to inform the organization of the result;
- establish if an affected worker has been in close contact with other workers or clients performing work activities without physical distancing in a team or pair, performing close contact services and inform those workers or clients of possible exposure to COVID-19, maintaining confidentiality as to the source of the potential exposure, and support affected workers to self-isolate or quarantine immediately;
- Retain details of other workers who have been in contact with affected workers in case CORONAVIRUS DISEASE is confirmed and there is a wider requirement to self-isolate;

- make sure the areas the affected person has been in are either isolated or cleaned and disinfected as soon as possible, giving particular attention to equipment, frequently touched surfaces (e.g. door handles, buttons for lifts) and common areas such as toilets;
- make sure that workers performing the cleaning or disinfection of affected areas are using appropriate PPE and following agreed safe working operating practices, based on assessment of the risks;
- Inform health authorities, e.g. if two or more confirmed cases of CORONAVIRUS DISEASE are connected to the workplace;
- Provide clear guidance on when it is safe for a worker who has had CORONAVIRUS DISEASE to return to the workplace;
- Provide information on measures that can be taken to facilitate return to work, ongoing support and rehabilitation, as appropriate.

**4.2 Managing illness of workers at home or in mobile settings**

The organization should establish a process for managing workers who develop symptoms of CORONAVIRUS DISEASE while working at home or in a mobile role. The organization should make sure that:

- workers are encouraged to report symptoms to the organization immediately;
- the affected person leaves the workplace, if this outside of their own home, using a safe method of transport avoiding public transport if possible, to a suitable safe place like home or a medical facility;

- workers are aware of, and directed to follow, regulations relating to self-isolation or quarantine (including if workers have been in close or prolonged contact with someone who has CORONAIRUS DISEASE.
- There is regular communication with the affected worker, to determine if symptoms develop further and/or the worker becomes seriously unwell;
- workers understand whether they should continue to perform work activities from home, if they are well enough, or if the time should be taken as sick leave;
- Workers understand the process for returning to work activities following self-isolation or recovering from CORONAVIRUS DISEASE.
- Reasonable adjustments are made, if necessary, to support a worker returning to work activities after contracting disease, taking into account both physical and psychological needs.

### 4.3 Managing Testing, contact tracing and quarantine activities

The organization should take action to make sure it is fully aware of current legislation or guidance from relevant regulators and health authorities on testing, contact tracing and quarantine.

In addition, the organizations should:

- encourage workers with symptoms to request a test at the earliest opportunity;

- encourage regular testing for workers who have extended interaction with other people as a result of their role, including workers with no symptoms;
- encourage the use of apps and research sites which monitor health and symptoms;
- support contact tracing by ensuring details of workers or people visiting the organization are maintained, as far as is practicable, and confidentiality is respected;
- require workers and other relevant interested parties to quarantine where this is required, due to:
    1. travel restrictions;
    2. advice from contact tracers, health authorities, or information received through apps or other communications;
- consider individual needs and circumstances if work-related activities can lead to the need to quarantine, whether at home or in another location, and support the cost of quarantine where appropriate;
- make reasonable adjustments for workers required to quarantine due to non-work-related activities and enable workers to take annual, special or unpaid leave, if appropriate;
- make its personal travel policy during the pandemic clear to all workers.

### 4.4 Impact on Psychological health and well-being

The organization should establish processes to manage the impact of the pandemic on workers' psychological health and well-being.

Psychological health and well-being can be affected by psychosocial hazards such as:

- uncertainty about what is expected, how long arrangements can last, impact on pay or working hours
- workload and work pace like too much or too little work, expectations of meeting short deadlines even if activities take longer due to amended ways of working etc;
- working hours such as. unpredictable hours, reduced or extended hours, new shift patterns ;
- role ambiguity like changes to what is expected from a role, new roles, lack of clarity ;
- lack of control resulting from rapid changes in risk levels, leading to sudden enforcement or easing of restrictions or amended ways of working ;
- lack of social support resulting from loneliness, physical isolation, communication issues
- impacts of prolonged isolation and remote working resulting from overexposure to screens, tiredness, boredom, lack of concentration, insomnia ;
- Job insecurity like. concern about possible job loss, domestic financial issues ;
    - difficulty in balancing work and home life

- specific roles that are higher risk due to frequent, close or prolonged interaction with other people ;
- worker's specific circumstances.

To manage risks to psychological health and well-being related to CORONAVIRUS DISEASE, initiate following actions as appropriate;

- promote a culture of trust, care and support by acknowledging that individual workers experience different issues and that anxieties or difficulties are valid and respected;
- enable regular confidential meetings ,remote or physical, as appropriate, to discuss issues and anxieties and to agree ways to support the worker;
- hold regular remote or physical meetings with teams of workers;
- allow flexible work hours and time off;
- assist workers in setting healthy boundaries between work and non-work time by communicating when they are expected to be working and available, taking into account the need for flexibility;
- allow workers more control over work pace and deadlines, if possible;
- give regular, clear and accurate information about the current situation in the organization and planned changes that can affect workers;

- consider providing appropriate PPE, masks, face coverings and other control measures for workers with concerns about being in the physical workplace, even if it is not required by the organization;

- offer additional resources to assist workers with managing their own psychological health and well-being like online programmes, websites, access to professionals offering bereavement and trauma counseling, financial advice etc..

**4.5 impact on different groups**

The organization should make sure that actions taken to manage risks arising from CORONAVIRUS DISEASE to work-related health, safety and well-being take into account the impacts on different groups of workers and other relevant interested parties.

The organization should, for example:

- make sure that issues and anxieties raised are respected and requests are accommodated as far as practicable;

- Continue to support working from home for workers who can effectively perform work activities at home and who are anxious about returning to the physical workplace;

- raise awareness and provide training to workers in order to meet the needs of people with disabilities such as providing access to suitable toilets, understanding how support animals operate, taking action to reduce communication difficulties caused by masks or face coverings etc.

- make sure facilities for faith groups are safely accessible;
- adapt roles and activities to reduce risks to vulnerable workers, if possible;
- make sure that communications, including electronic communications, are accessible like websites, online appointments or ordering systems;

**4.6 Resources for managing risks**

The organization should determine what resources are needed to effectively manage the risks related to CORONAVIRUS DISEASE and make sure sufficient resources are in place. The organization should establish processes to help make sure that essential resources are maintained, appropriately managed and can be supplied reliably as needed.

Workers with responsibility for managing resources to mitigate the risks related to CORONAVIRUS DISEASE should be clearly identified, and this should be communicated to all workers and other relevant interested parties. The organization should make sure that there is a process to enable ongoing dialogue with workers about specific needs for resources to manage risks related to CORONAVIRUS DISEASE and how workers can escalate issues.

When determining the resources needed to start, resume and maintain essential activities, the organization should consider:

- human resources, including practical and psychological support to workers, and processes to manage reduced human resources due to illness or self-isolation;
- financial resources;
- appropriate PPE, including specific provision for workers with cleaning and disinfection roles;

    - hand washing, hand sanitizing, and cleaning and disinfection materials;
    - adequate and safe provision of toilet facilities;
    - technology;
- infrastructure and equipment relating to waste, water and energy management
    - communication means
    - the need for, and availability of, additional training to make sure workers are competent to take on additional roles or activities.

The organization should make sure that temporary, prolonged or permanent absence of workers (e.g. through sickness, self-isolation or quarantine, job losses) does not put the health or safety of available workers at risk. The organization should make sure that workers are competent to perform roles or activities they are required to perform, particularly if workers are expected to take on new tasks.

The organization should take actions to minimize additional workload and make sure that any additional workload is only short term. Line managers should monitor workload and the impact on affected workers so that individual workers do not work beyond agreed working hours and take rest periods and time off work.

# 5. COMMUNICATION

## 5.1 General

The organization should communicate its commitment to managing the risks related to CORONAVIRUS DISEASE and inform workers and other relevant interested parties of:

- general safety measures and controls;
- required ways of working, taking into account the needs of individuals and groups of workers;
- what is expected of them;
- what they can expect from the organization;
- how to report concerns or safety incidents.

The organization should make sure that there is regular communication from top management to workers at all levels, to demonstrate commitment to policies and agreed ways of working during the pandemic.

The organization should use a combination of formal and informal communication methods (e.g. intranet, website, emails, signs, images, symbols, phone calls, audio announcements, videos) so messages are accessible and can be understood by all relevant interested parties, including people with disabilities, non-native speakers and people with differing levels of literacy. The organization should make sure that standardized symbols are used, wherever possible, to avoid misinterpretation. Preferred methods of communication like. emails or personal phone calls, rather than video conferences with groups) should be taken into account for workers with different needs, including making adjustments for neurodiversity Communication with workers and other relevant interested parties should be two-way and methods should facilitate ongoing conversation as well as more formal consultation.

Communications should provide clear and up-to-date guidance on physical distancing, hygiene and required behaviors:

- before arrival at the workplace by phone, website, intranet, email ;
- on arrival at the workplace (e.g. signs, posters, screens, announcements);
- at first entry into a workplace
- throughout the workplace by displaying signs, posters, screens, announcements etc.

Communications should also provide clear guidance on facilities and functions that are or are not available such as canteens, fridges, shared equipment, first aid, HR, IT etc. Regular communications should be provided on changes to processes, guidance and the levels of risk related to CORONAVIRUS DISEASE,

The organization should:

- establish who is responsible for communicating safety guidance to visitors, delivery workers, customers and other people ensuring more than one person is trained to perform this role);

- make sure THAT communications are accessible and useable by all workers and relevant interested parties, including contractors and agency workers;

- provide necessary training to workers who act as hosts for visitors, or who need to interact with delivery workers, customers, the public, etc.;
- communicate relevant information about operational changes, safety measures and controls to suppliers, customers and other relevant interested parties;
- review communications frequently to make sure they are current and effective and take action if issues are identified;
- establish effective day-to-day communication mechanisms in workplaces to enable compliance with physical distance requirements, including where noise levels are high and cannot be reduced.

**5.2** For **first entry into a workplace**

The organization should take all reasonable measures so that workers and other relevant interested parties understand the behaviours, processes and working practices required to manage the risk of transmission of CORONAVIRUS DISEASE before entering a workplace for the first time or returning from absence from the workplace. In addition to the actions recommended above organization should:

- develop communication and training materials and deliver training as required through video training or electronic methods;
- provide guidance on safe travel to and from work like encouraging walking, cycling and personal vehicles where possible, and

physical distancing and masks or face coverings if workers need to use public transport;

- provide clear guidance on staggered start and finish times, flexible working hours, shifts or any other altered working patterns or schedules;
- provide guidance on physical distancing, hygiene and general ways of working;
- Communicate new processes for entering the workplace, beginning work and the use of common areas like. lifts/elevators, stairways, toilets, kitchens, corridors etc;
- communicate guidance on safe interaction with visitors, customers, service users and other people;
- communicate changes to emergency procedures

**5.3 Ongoing communication**

The organization should make sure that all workers are regularly reminded of safety measures and controls and that they are kept up to date if these are changed or additional safety measures or controls are implemented.

The organization should:

a) make sure ongoing engagement with workers and worker representatives, where they exist, and take actions to understand any unforeseen impacts of changes to ways of working, how work is organized and workplaces

b) communicate regularly with workers, including those working remotely, to check physical and psychological health and well-being, and to give clear information on issues that are known to negatively affect psychological health

**5.4 health and hygienic conditions**

The organization should implement processes to keep the workplace clean, to reduce the risk of transmission of CORONAVIRUS DISEASE from contaminated surfaces, and to enable good hygiene throughout working hours and at the end of each working shift. The organization should make sure that workers are made aware of the importance of frequent and effective hand washing to limit transmission of CORONAVIRUS DISEASE. The organization should communicate to workers that:

- hands should be washed with clean and hot water and soap for minimum 20 seconds
- hands should be sanitized with a hand sanitiser suitable for safe and effective use against CORONAVIRUS DISEASE. Use sanitiser containing a minimum of 60 % ethanol or 70 % isopropyl alcohol, if hand washing is not possible;
- visibly soiled hands should be washed before using hand sanitizer, if possible.

The organization should make sure that hand sanitizers conform to relevant standards for the type and concentration of alcohol on labels and be aware of the possibility of counterfeit, low quality or incorrectly formulated products on the market.

The organization should implement processes to ensure:

- workers are encouraged to wash their hands or sanitize at frequent intervals, and communicate when this should be done before entering or leaving an area of the workplace, before and after breaks, before and after handling shared resources such as telephones, computers, tools, drink dispensers, before and after using common areas ;
- additional hand washing and/or hand sanitizing facilities are available in places where workers are present or move through. entrances, exits, near elevators, common areas, operational areas ;
- additional materials are available to workers to enable frequent cleaning and disinfection of workstations and equipment, including between use by different workers;
- frequent cleaning and disinfection of surfaces that are touched regularly like door handles, light switches, counters, pay points, testing surfaces, lift/elevator controls, shared resources etc;
- effective, adequate and frequent waste disposal, including separate, secure waste disposal for single-use PPE and disposable masks and face coverings;
- promotion of good hygiene practices, including posters and signs to remind workers of required hand washing techniques and frequency, the need to avoid touching faces, and to cough or sneeze into a disposable tissue or into their elbow;
- safe use of toilets, including increased ventilation, enhanced and more frequent cleaning and disinfection, encouraging use of paper towels and managing use to reduce crowding

- safe use of showers and changing rooms, designating specific facilities for small groups where this is possible.

To avoid transmission from contamination of surfaces, the organization should implement fixed workstations, zones, desks and/or equipment, and require workers to keep personal belongings in personal spaces, such as lockers or bags, ensuring belongings are removed from the workplace at the end of each shift.

The organization should take action to reduce the risk of transmission of CORONAVIRUS DISEASE through contact with objects that come into the workplace and vehicles used by the organization. The organization should:

- restrict non-essential deliveries, including personal deliveries to workers;
- clean and disinfect materials, equipment and other objects entering the workplace;
- clean and disinfect touch points of shared equipment after each use;
- regularly clean and disinfect vehicles used for work activities, including vehicles workers drive home;
- increase frequency of hand washing for workers handling deliveries or provide hand sanitizer where this is not practical.

# 6. USE OF PERSONAL PROTECTIVE EQUIPMENT

## 6.1 General

PPE protects the user against health or safety risks at work. In the context of COVID-19, PPE such as respiratory equipment and face shields (when used with a mask) can be used. If workers are required to use PPE to protect against risks unrelated to transmission of CORONAVIRUS DISEASE they should continue to do so.

There is increasing evidence that masks and face coverings, including homemade textile face coverings, provide some protection against the transmission of CORONAVIRUS DISEASE by capturing droplets released through breathing, coughing, sneezing and talking. Face coverings, used in conjunction with physical distancing, hand washing and other hygiene measures (see Clause 10) are an effective measure in reducing the risks related to COVID-19.

Specialist PPE and medical devices (e.g. respirators, masks to protect workers from dust and other industrial airborne hazards) should be reserved for those who need them to perform their roles.

The organization should take into account situations where temporary removal of PPE, masks and/or face coverings can be required or where workers or other interested parties have specific needs. These can include:

- temporary removal of masks or face coverings for identification or other security purposes;
- interaction with workers and other interested parties with hearing impairments who lip read.

If temporary removal of PPE, masks and/or face coverings is necessary, physical distancing should be ensured. Hand washing (or sanitization) should also be ensured to avoid cross-contamination when putting on or taking off PPE, masks or face coverings. To improve communication for people who lip read and for other interested parties, the organization should facilitate use of appropriate transparent face shields, if this is possible.

If additional PPE, masks or face coverings are required to manage the risks related to CORONAVIRUS DISEASE, the organization should:

- establish guidelines for when and how PPE, masks and/or face coverings should be used and provide training if necessary;
- provide suitable PPE and/or masks free of charge;
- make sure PPE and masks are correctly fitted, and instruct workers on appropriate use and safe disposal after use;
- encourage workers to take regular breaks to minimize fatigue caused by using PPE, which can lead to reduced compliance with safety measures and unsafe use of equipment;
- clean, disinfect or launder contaminated reusable PPE.

The organization should support workers who choose to use a mask or face covering not required by the organization

The organization should advise workers to:

- wash their hands or use hand sanitizer before putting the mask or face covering on and after removing it ;
- continue to regularly wash hands, or sanitize hands if this is not possible;
- avoid touching their face or mask/face covering, to avoid contamination;
- change their mask or face covering if it becomes damp, or if it has been touched with dirty or potentially contaminated hands;
- change their mask or face covering each day, as a minimum
- dispose of or store masks or face coverings in a sealed container if removed, to avoid contamination of other surfaces;
- wash reusable masks or face coverings at a high temperature before/after each use if the material is washable;
- securely dispose of masks or face coverings after single use if the material is not washable;
- continue to comply with physical distancing guidelines, wherever possible.

# 7. PROCESSESS

## 7.1 General

The organization should make sure that processes are in place to address the risks identified including implementing measures to enable home working, physical distancing, and other safety measures and controls in the workplace.

The organization should assess if the measures introduced negatively impact existing security measures or introduce new security risks, and take actions to address these risks.

The organization should take measures to reduce background noise in the workplace as far as practicable by lowering music, reducing the time that devices such as hairdryers are used, to reduce the need for people to raise their voices. Raised voices, including shouting, singing and other types of voice projection, can increase the range of droplet transmission. Noise reduction, where practicable, is therefore important both in places where people are using masks or face coverings, which can muffle sound, and in situations where physical distancing is difficult or impossible like close contact roles, such as hairdressers, tattooists, physical therapists, or social settings, such as pubs and restaurants.

In activities and situations where it is impossible to fully comply with physical distancing guidelines, the organization should implement the actions and make sure that activity times involved are kept as short as possible.

If an activity requires close contact work for a sustained period without being able to comply with physical distancing guidelines or bringing workers into contact with people other than their assigned team or pair, the organization should assess if the activity can go ahead safely.

No worker should be obliged to work in an unsafe work environment.

### 7.2 First return to a workplace

The organization should develop a process to communicate changes to the workplace and ways of working to all workers on first arrival or return to a workplace and it should make sure that this is regularly reviewed and updated as circumstances change. This should be in addition to communications provided before the return to work and should include guidance for specific roles or activities.

The organization should:

- make sure that all workers returning to the workplace, or attending a different workplace or site, are provided with full instructions and information on arrival;
- Communicate information about potential hazards that can arise if there are reduced numbers of workers;
- Limit the number of workers being given instruction about first entry to the workplace at one time to enable physical distancing;
- Consider using outside spaces for instructions on first entry, where safe and possible.

The organization should raise awareness of CORONAVIRUS DISEASE symptoms and establish appropriate processes for health screening of workers and other people like visitors, service users, and prior to anyone entering the workplace. This can include self-reporting and/or temperature checks.

Advice and recommendations can be provided by occupational health professionals, either through the organization's internal resources or through consultation with external services or professional bodies.

The organization should take following precautionary majors, as appropriate, while employee enter and leave the workplace

- Stagger arrival and departure times to reduce crowding at entry and exit points;
- Provide additional entry and exit points if possible;
- Provide additional parking or facilities, such as bike racks, where possible;
- Limit the number of passengers in vehicles used by the organization, such as minibuses
- Use physical distancing indicators on the floors or walls and introduce one-way systems at entry and exit points, if possible;

- Create separate entry and exit points for high-risk work areas or sites
- make sure touch-based security devices, such as keypads, biometric readers and electronic pass points, are regularly sanitized and raise awareness that no physical contact is needed between access cards and readers;
- make sure safety measures introduced to manage the risks related to CORONAVIRUS DISEASE do not unintentionally create security risks ;
- Provide storage for workers' and service users' clothes and bags, preferably storage dedicated to single-person use;
- Provide facilities for workers to change into work clothing and equipment on-site, where physical distancing and hygiene guidelines can be met;
- Clean, disinfect or wash clothing and equipment like uniforms, hard hats, goggles, gloves, on-site if possible.

### 7.3 Movement around and between workplaces

Maintain physical distancing guidelines wherever possible, while people move through the workplace and between workplaces. In order to manage this, carry out the following activities.

- Reduce movement within buildings and sites if possible, restrict access to specific work areas to workers who need to be there, encourage use of radios or telephones, where permitted, cleaning them between use if these are shared;
- Enable no-contact access controls in areas where controlled entry is necessary like automated doors;
- Remove access controls that need to be touched like electronic barriers, keypads etc, in low-risk areas, to reduce surface contamination;

- Install barriers to avoid contact between workers performing health screening and the person who is being screened at entrances, between workplaces and in any other location where health screening takes place ;

- Use teams/pairs or timed booking processes to reduce the number of people in a work area at one time;
- Introduce one-way systems through buildings, paying particular attention to long or narrow corridors, stairways, walkways and turnstiles;
- encourage the use of stairways and reducing maximum occupancy for lifts/elevators, providing hand sanitizers for their operation, and

ensuring regular cleaning and disinfection of commonly touched areas;

- Enable people with disabilities to safely access and use lifts/elevators.

**7.4 Work zones and workstations**

The organization should make sure that physical distancing guidelines can be maintained between individual workers in work zones and at workstations, wherever possible.

To facilitate safe working practices, the organization should:

- review work zones and, where possible, move workstations to enable physical distancing between each station, paying attention to the space needed to safely move to and from workstations, if this involves passing other workers;
- Arrange workstations so that workers are side-by-side, back-to-back or diagonal to each other, rather than face-to-face;
- Consider blocking the use of some workstations, or use screens to separate workers if workstations are fixed at less than the recommended distance;
- Assign workstations and equipment to individual workers, wherever possible, or teams/pairs where this is not possible
- use floor or wall markers to indicate recommended physical distancing guidelines;

- put in place physical barriers to enforce physical distancing to the extent possible, where it is safe to do so without introducing new OH&S or other risks or negatively impacting people with disabilities;
- Reduce the number of workers in a work zone to enable physical distancing in restricted spaces;
- Limit the use of high-touch items and shared equipment, and enable frequent cleaning and disinfection.

### 7.5 Safe Use of common areas

The organization should implement processes to facilitate the safe use of essential common areas, including, as a minimum:

- frequent cleaning and disinfection, including between uses by different groups of people;
- limiting the number of people in common areas at one time;
- limiting how long people can be in common areas;
- physical distancing.

The organization should also consider, as appropriate:

- staggering the times when workers arrive or leave when working with other organizations in shared spaces, to reduce crowding in common areas such as lifts/elevators, reception, corridors and security points;

- staggering break times and encouraging the use of safe outside areas, if possible;
- encouraging the use of outdoor spaces for work activities, where practical;
- creating additional common spaces in other parts of the workplace;
- installing screens to protect workers in reception or similar areas;
- encouraging workers to bring in their own food, or providing packaged meals to avoid opening canteens, where appropriate;
- avoiding the use of shared resources, such as cups, plates and spoons, and ensuring water taps and drinks dispensers are cleaned or disinfected by the user after each use;
- moving seating and tables to enable physical distancing and reduce face-to-face interaction;
- encouraging workers to remain in the workplace including designated outdoor space. throughout working hours, and requiring compliance with physical distancing guidelines if leaving the workplace;
- regulating the use of locker or changing rooms, showers and other common facilities (e.g. baby and family rooms, faith rooms and associated foot-wash facilities);
- encouraging storage of personal items in personal spaces, e.g. lockers, during working hours.

**7.6 Use of wash rooms**

The organization should consider additional measures to facilitate the safe use of toilet facilities by workers and other interested parties, including those provided for people with disabilities. Actions can include:

- managing the use of toilet facilities to avoid crowding;
- establishing more frequent and enhanced cleaning and disinfection including touch points such as toilet seats, locks, flushes, grab rails and hoists and waste disposal;
- using signage to direct users to the nearest available toilet, if toilets are temporarily closed for in-depth cleaning;
- limiting the number of cubicles and urinals available in a block of toilets, to promote physical distancing;
- using signage to encourage users to close toilet lids before flushing, where lids are fitted;
- ensuring a system is in place to allow queues for toilets to form outside of the facility, rather than in the confined space;
- requesting workers or visitors to use a single designated set of facilities within a workplace, taking into account users with special needs;
- providing paper towels, and ensuring levels of paper towels are monitored and maintained and that there is frequent, safe disposal of waste;

- using automatic and foot-operated equipment, rather than manual equipment like sensor operated taps, soap dispensers, flushes, foot-operated bins etc.
- increasing the monitoring and replenishment of supplies like soap, sanitizer, paper towels, toilet paper etc.

### 7.7 Managing Meetings and visits to the workplace

The organization should limit visits to the physical workplace and use remote working technology to minimize both external and internal face-to-face meetings, particularly while restrictions are in place.

If face-to-face meetings or visitors to the workplace are essential, the organization should communicate expected behaviors and processes for safely entering the building in advance of the visit, including health screening and self-reporting health status.

The organization should:

- Restrict access to required visitors only;
- Take into account where visitors are travelling from and if additional safety measures are needed;
- Limit the number of visitors in the workplace at any one time;
- Limit visits to specific times;
- Provide separate toilet facilities for visitors, if possible;
- Revise schedules for essential service and other contractor visits to reduce interaction

- record visitor details to enable contact tracing by recording. names, dates, who is hosting the visit, names of other people in the workplace or through work activities the visitor has close or prolonged contact with , taking measures to make sure these data are protected and destroyed after an agreed period of time ,not less than 14 days or following official guidance;
- Revise how visitor details are recorded and how visitors enter and exit the workplace by verifying details recorded by a receptionist to avoid shared pens, using one-way systems to enter and exit, using disposable visitor badges;
- require visitors to comply with physical distancing guidelines and other safety measures and controls;

- make sure reasonable adjustments are made for people with disabilities who have access requirements and are attending meetings.

If physical meetings are essential, the organization should:

- Limit participation to the minimum number of essential people and maintain physical distancing guidelines;
- Avoid shared resources like. pens, water or coffee jugsetc;
- Provide hand sanitizer in the meeting room;
- Hold meetings outside or in well-ventilated rooms, if possible;
- Use floor or wall markings to indicate acceptable physical distancing guidelines.

## 7.8 Precautions while working with the public

The organization should make sure that controls are in place to maintain physical distancing and the organization should take following actions as and where required to minimize risks of infection to and from workers through interaction with the public including customers, clients, service users and other people in both indoor and outdoor workplaces.

- Training workers with public facing roles to be aware of how to communicate safety measures to members of the public, including people with disabilities who have individual needs

- Using posters, signs, marketing emails and other communications to inform members of the public of safety measures and controls and how to maintain physical distancing;

- Making regular announcements to remind members of the public to maintain physical distancing and follow other safety measures;

- Limiting the number of members of the public in a building or confined outdoor space so that physical distancing can be maintained;

- using safe outdoor spaces for queuing, where possible, using floor or wall markings to indicate physical distancing intervals, ensuring queues do not cause additional safety hazards and that street furniture is not removed, causing additional security risks

- Providing hand sanitizer at entrances and exits to buildings and outdoor spaces, and other areas of outdoor spaces where there is potential risk of transmission;

- Monitoring the use of masks or face coverings where this is mandatory;
- considering provision of disposable masks for customers, clients or service users, and other members of the public who do not have their own or who are wearing unsuitable masks or face coverings;
- ensuring cleaning and disinfection of frequently touched areas and shared resources, such as card payment and cash machine keypads, sales counters and bars, handles of baskets and trolleys, treatment beds or chairs, gym equipment etc;

- Limiting handling of products through different display methods, signs, rotation of high-touch items;

    - Providing physical barriers, such as screens, in places where interaction between workers and members of the public is frequent like pay points, customer service desks etc.
    - Reducing non-essential public facilities if physical distancing cannot be complied
    - limiting time spent in close contact with customers or service users, adapting services as necessary (e.g. ensuring hair and beauty treatments are time-limited; using electronic devices for ordering food and drink; using designated pairs of workers to carry heavy items to customers' vehicles, rather than a single worker assisting a customer to carry the item);
    - Providing well-indicated toilet facilities, with physical distancing marked for queues and a suitably trained worker in attendance in

busy facilities to regulate entry and make sure enhanced cleaning, waste disposal and replenishment of supplies;

- Encouraging contactless payment and refunds;
- Establishing no-contact collection and return points;
- Staggering collection times;
- Establishing a booking system, if appropriate

**7.9 Work-related travel**

**Consider following if work related travel is must,**

- take into account the different forms of travel required to complete a journey and the places workers are required to transit through railway stations, airports, hotels ;
- Take into account the varying requirements of different travel organizations and hubs like airline or ferry restrictions, specific requirements for airports or ports etc;
- Encourage flexibility of travel times to avoid peak times on public transport;
- Encourage people to cycle, use electric bicycles or scooters, or their own vehicle, where practicable;
- Determine locations of essential facilities like. toilets, food and drink and give guidance on safe use;
- Centrally log if a worker is required to stay away from home overnight, and make sure overnight accommodation complies with physical distancing and hygiene guidelines.

### 7.10 Precautions while travelling on roads and in vehicles

- Keep two (2) meters of physical distancing between two persons
- Minimize the number of people travelling together in any one vehicle;
- Use fixed teams or pairs while travelling;
- Open windows to increase ventilation in motor vehicles, where practicable;

- make sure vehicles are cleaned and disinfected between shifts and before use by other workers;

- Require workers to avoid sitting face-to-face;
- Encourage the use of masks or face coverings if more than one person is in a vehicle, including in taxis.

### 7.11 Safe Shipment and receipt of deliveries

Make sure the following, in order to safe shipment and receipt of deliveries of all documents and other items

- Minimize person-to-person contact during deliveries, including during payment and exchange of documentation
- provide guidance and training to workers taking deliveries at home, or in another location not controlled by the organization, on safe handling and distribution;

- Revise pick-up and drop-off collection points use zones with physical distancing markings, no-contact drop-offs to customers and other work sites wherever possible,
- Reduce the frequency of incoming deliveries use central procurement processes to avoid external deliveries to different sites, ordering larger quantities less often wherever possible
- Use single workers or fixed pairs to load or unload vehicles;
- Provide controlled, safe access to welfare facilities for delivery drivers and outsiders,

- Encourage drivers and outsiders to stay in their vehicles where this does not compromise safe working practice;
- make sure regular cleaning and disinfection of reusable delivery boxes, loading equipment, etc.;
- Consider cleaning or disinfection of delivered items, or isolate items that cannot be disinfected, following official guidance for different materials, to allow for natural decay of the CORONAVIRUS DISEASE virus on surfaces.

# 8. PERFORMANCE EVALUATION, MANAGEMENT REVIEW, INCIDENTS AND REPORTING

## 8.1 Monitoring and evaluation

The organization should use a systematic approach to monitor and evaluate:

- how effectively safety measures and controls protect workers;
- how the work is being done;
- compliance with safety measures in the workplace;
- the rate of infection among workers;
- levels of worker absence and the impact on available workers;
- changes in community risk levels or other external issues

Monitoring and evaluation activities should:

- determine the extent to which the guidance is being complied with;
- determine if processes for ongoing assessments of risks are in place and operating effectively;
- determine the extent to which controls are working and if these need to be changed, enhanced or enforced more actively;
- determine if the use of controls is creating new risks (of any type) that need to be addressed;

- take into account feedback from workers and worker representatives, where they exist, and other interested parties (e.g. customers, service users).

The organization should consider implementing increased supervision of activities to make sure safety measures are complied with. The organization should review the outputs of monitoring and evaluation at regular intervals and take into account:

- issues identified with levels of compliance with safety measures and controls;
- incidents reported by workers and other relevant interested parties;
- the root cause(s) of incidents;
- the effectiveness of actions taken to deal with incidents, including actions taken at the time of the incident and actions to address the root cause(s) of the incident.

The results of the management review should be communicated to workers and other relevant interested parties, as appropriate. Communications should include actions taken and other improvement measures that are, or will be, introduced (see Clause 14).

Reporting to external interested parties

If a worker contracts CORONAVIRUS DISEASE due to work-related exposure to the disease, it should be reported to the appropriate regulator or health authority.

The organization should be aware that reporting requirements can change as circumstances change. The organization should regularly review reporting requirements and make sure information is up to date.

When deciding if a report is required, the organization should determine if there is reasonable evidence that work-related exposure, rather than general social exposure, is the likely cause of the disease.

Factors to take into account when determining if contracting CORONAVIRUS DISEASE has been caused by work-related exposure include:

- if the nature of work activities or work organization has increased the risk of workers becoming exposed;
- any specific, identifiable incident that led to an increased risk of exposure;
- if work activities directly brought a worker into contact with a known corona virus hazard without effective control measures being used such as physical distancing, PPE.

If more than one worker contracts disease, regardless of whether this is caused by work-related exposure or not, the organization should report this to the relevant regulators or health authorities, so that actions can be considered or implemented to control an outbreak and prevent further CORONAVIRUS DISEASE cases in the organization or community.

# 9. IMPROVEMENT

## 9.1 General

The organization should determine opportunities for improving how it manages risks related to CORONAVIRUS DISEASE and implement necessary actions. This includes staying informed about the status of CORONAVIRUS DISEASE cases, new information on the disease, and updates on infection controls and treatment.

The organization should take into account the results of monitoring, evaluation and review

- take immediate actions to improve or change safety measures and controls that are not effective;
- implement additional safety measures and controls if needed, taking into account the security implications of any new measures introduced;
- address changes to the external and internal issues that can affect work-related health, safety and well-being , including changes to local, regional or national risk levels, official guidance or legal requirements;
- encourage ongoing consultation and participation of workers and worker representatives, where they exist, during the monitoring, evaluation and review, and address their concerns.

To make sure the organization continues to manage the risks related to CORONA VIRUS DISEASE, it should review the recommendations in this document regularly, to take into account the dynamic nature of the situation.

### 9.2 Protective security considerations

Safety measures

Safety Operations and practices

Normal protective security operations and practices should be taken into account when implementing CORONAVIRUS DISEASE related measures or controls.

The organization should:

- consult with and involve their security department, where this exists, in the implementation of proposed safety measures;
- consult with security workers and take into account the security arrangements of partner organizations and organizations sharing facilities;
- take security into account throughout all revised risk assessments;

- Make sure workers with responsibility for implementing measures to manage the risks from CORONAVIRUS DISEASE consult with workers in security roles, and coordinate and clarify roles and responsibilities.

### 9.3 Security Measures

Protective security measures should not be removed, altered or reduced without undertaking a security risk assessment. Where necessary, the organization should seek advice from relevant protective security experts (e.g. from the national security authority or police counter-terrorism specialists).

The organization should take into account measures not primarily intended for protective security, but which provide a security benefit (e.g. removal of street furniture can make moving or queuing pedestrians more vulnerable to vehicle-as-a-weapon attacks).

The organization should:

- make sure that security workers remain focused on security duties;
- make sure additional resources are provided if CORONAVIRUS DISEASE safety measures create the need for additional workers for supervision or other activities
- confirm that security workers feel safe to perform their duties and they have access to appropriate PPE and hand-washing facilities ;
- provide guidance on how to perform security duties without significantly increasing risks to personal health and safety , guidance on physical distancing where people are asked to remove masks or face coverings for identification purposes ;
- agree a method for security workers to raise concerns.

**9.4 CORONA VIRUS DISEASE safety measures.**

The implementation of additional measures to manage work-related risk from CORONAVIRUS DISEASE can have a disproportionately negative impact on people with a disability.

This annex provides further considerations for organizations to make sure CORONAVIRUS DISEASE measures do not exclude people or create additional unintended risks.

Individual needs to consider

The organization should encourage discussion and engagement with workers and worker representatives, where they exist, to make sure individual needs are understood. The organization should take into account that:

- not all people with a disability are more vulnerable to CORONAVIRUS DISEASE;
- many people have vulnerabilities to CORONAVIRUS DISEASE that are not visibly apparent like diabetes, respiratory conditions, heart conditions etc;
- many other disabilities are also not visibly apparent, and adjustments can be necessary to meet individual needs.

In addition to the general measures mentioned in these guidelines, the organization should consider surveying all workers in order to understand recent and ongoing health, safety and well-being issues and personal circumstances.

Factors to consider

The organization should also take into account:

Factors affecting the outside of a workplace, including:

- maintaining existing parking facilities for people with a disability and not reducing these facilities when creating additional space for customers to queue ;
- creating safe "drop off" zones for people who are at higher risk from CORONAVIRUS DISEASE or with underlying health conditions ;
- ensuring there is sufficient space including consideration of physical distancing requirements for wheelchair and other

mobility aid users when creating new, one-way or split routes around workplaces;

- ensuring alternative routes are provided if new routes are not step free;
- ensuring a detectable warning surface is provided when changes such as creation of additional cycle stands are implemented;

Factors affecting the inside of a workplace, including:

- ensuring hand sanitizer is accessible to all
- recognizing that one-way systems can create longer routes, which affect people with mobility impairments
- ensuring there is sufficient space including consideration of physical distancing requirements) for wheelchair and other mobility aid users when creating new, one-way or split routes within buildings;
- enabling workers who require a care taker or assistant to book side-by-side workstations or desks;
- designating specific accessible toilet facilities for people considered to be at higher risk of contracting CORONAVIRUS DISEASE or getting severe illness from COVID-19, and implementing additional and more frequent cleaning and disinfection to make sure required hygiene standards are consistently met;

- enabling workers considered to be at higher risk of contracting CORONAVIRUS DISEASE or getting severe illness from CORONAVIRUS DISEASE to work together, to facilitate higher levels of physical distancing and hygiene and limit interaction with other people;

- factors relating to communication, including:

    - the communication needs of people who are blind, visually impaired or deaf;

    - ensuring signs and notices use clear, consistent and simple language and recognized symbols, and are large enough;

    - considering the use of closed caption subtitles on videos;

    - considering the creation of a video demonstrating changes and providing induction to the workplace that workers and other relevant interested parties can access before entering the workplace;

    - recognizing that masks and face coverings create communication issues for people who rely on lip reading and perception of emotion through facial expressions, and enabling additional measures to be used if possible (e.g. transparent face shields used with physical distancing to enable masks and face coverings to be removed for direct communication);

# References –

1. acequalityconsultants.com

2. Basics of occupational, health and safety management system – Rajendra Patil, KDP – AMAZON

3. Guide to implement Quality Management System- ISO 9001-2015 – R S Patil, KDP – AMAZON

4. Quality news and views – Monthly news letters

# OTHER BOOKS BY THE AUTHOR

1. Quality management system( Marathi) – Phadke publications, Kolhapur – 2003,2020

2. Guide to implement Technical specifications ( ISO/TS 16949) –
   a. 2006 -- Phadke Publishers, Kolhapur

3. Performance improvement through five S - ACE QUALITY CONSULTANTS

   a. 2015 , 2017

4. Guide to implement Quality Management System- ISO 9001-2015

5. QUALITY and productivity improvement tools, 2019 – kdp -AMAZON

6. Quality system – IATF 16949 – simplified – kdp –AMAZON – 2020

7. Basics of quality management system.

8. Basics of occupational, health and safety management system – KDP - AMAZON

www.ingramcontent.com/pod-product-compliance
Ingram Content Group UK Ltd.
Pitfield, Milton Keynes, MK11 3LW, UK
UKHW022020190726
13853UKWH00005B/2029